**Jörg Matthias Determann** is Assistant Professor of History at Virginia Commonwealth University in Qatar. He is the author of *Researching Biology and Evolution in the Gulf States: Networks of Science in the Middle East* and *Historiography in Saudi Arabia: Globalization and the State in the Middle East* (both I.B.Tauris). He has been a Postdoctoral Fellow at Zentrum Moderner Orient and Freie Universität Berlin, and holds a PhD from the School of Oriental and African Studies (SOAS), University of London.

‘Exceptionally well researched, drawing on a copious array of archival and original source material in multiple languages, this book makes a significant contribution to the modern history of astronomy in the Middle East.’

**James Clay Moltz, Professor,**
**Naval Postgraduate School, Monterey, California**

‘Jörg Matthias Determann’s book shows that space science has been a formative element of Middle East modernity. States as disparate as Nasser’s Egypt and the Gulf monarchy of Qatar have looked to the stars to further nationalist as well as internationalist agendas, mobilising the notions of “golden age” and “renaissance” in the service of a distinctively modern politics of science, industry, and military technology. *Space Science and the Arab World* is an exciting addition to the emerging genre of scholarship on science in the modern Middle East.’

**Daniel A Stolz, Visiting Assistant Professor,**
**Northwestern University**

# SPACE SCIENCE AND THE ARAB WORLD

## Astronauts, Observatories and Nationalism in the Middle East

Jörg Matthias Determann

I.B. TAURIS
LONDON · NEW YORK

Published in 2018 by
I.B.Tauris & Co. Ltd
London • New York
www.ibtauris.com



Library of Modern Middle East Studies 215

ISBN: 978 1 78831 014 7
eISBN: 978 1 78672 352 9
ePDF: 978 1 78673 352 8

A full CIP record for this book is available from the British Library
A full CIP record is available from the Library of Congress

Library of Congress Catalog Card Number: available

Typeset in Stone Serif by OKS Prepress Services, Chennai, India

*To Jeanne, my star*

# Contents

# List of Figures

# Abbreviations

| | |
|---|---|
| Arabsat | Arab Satellite Communications Organization |
| ASTP | Apollo–Soyuz Test Project |
| AUASS | Arab Union for Astronomy and Space Sciences |
| AUB | American University of Beirut |
| COPUOS | Committee on the Peaceful Uses of Outer Space |
| CRAAG | Centre de Recherche en Astronomie, Astrophysique et Géophysique |
| EIAST | Emirates Institution for Advanced Science and Technology |
| ESA | European Space Agency |
| IAU | International Astronomical Union |
| KAU | King Abdulaziz University |
| KFUPM | King Fahd University of Petroleum and Minerals |
| NASA | National Aeronautics and Space Administration |
| NORT | Network of Oriental Robotic Telescopes |
| PLO | Palestine Liberation Organization |
| QF | Qatar Foundation |
| QNRF | Qatar National Research Fund |
| SPC | Syrian Protestant College |
| UAE | United Arab Emirates |
| UNESCO | United Nations Educational, Scientific and Cultural Organization |
| UNISPACE | United Nations Conference on the Exploration and Peaceful Uses of Outer Space |
| USAF | United States Air Force |
| USSR | Union of Soviet Socialist Republics |

# Acknowledgements

The transnational history of Arab space science accompanied me as I moved from Britain to Germany and Qatar. I developed the initial idea for this research project as a research student under the supervision of Konrad Hirschler at the School of Oriental and African Studies (SOAS), University of London. The idea stayed with me while I worked as a postdoctoral fellow under Ulrike Freitag and Gudrun Krämer at Zentrum Moderner Orient and the Berlin Graduate School Muslim Cultures and Societies. I undertook most of the research for this book, however, while serving as a faculty member at Virginia Commonwealth University in Qatar. I thank especially Akel Kahera, Allyson Vanstone, Byrad Yyelland, Craig Anz, Greet Provoost, Junita Patrick, Line Christiansen and Patty Paine for their generous support during this period.

At the various stages of writing the book, I received feedback from scholars of history, the Middle East and science. Imad Mansour and Sean Roberts provided great comments on the proposal. The following people kindly reviewed draft chapters: Anto Mohsin, Cliff Cunningham, Dan Stolz, Farouk El-Baz, Omer Can Aksoy, Radha Dalal, Robert Smith, Stefano Taglia and Trinidad Rico. Hala Auji commented on Chapter 2 and shared with me the minutes of the Syrian Protestant College's Board of Trustees. I also thank my students as well as Iman Khamis, James Roach, Khaled Saoud and Ryad Ghanam for their inspiring conversation about Arab scientists. After I.B.Tauris accepted my proposal, I was very fortunate to receive excellent feedback from my outstanding editor Sophie Rudland and three anonymous peer reviewers. I also thank Sophie's colleagues at I.B.Tauris, especially Lisa Goodrum and Angelique Neumann, for all their work in producing this book. Oliver Jarvis from Global Proofreading and Copy-editing corrected my English.

While writing this book, I was fortunate to receive a lot of feedback from audiences in the Middle East, Europe and North America between 2014 and 2016. I thank Karl Widerquist, Leah Long, Mohanalakshmi Rajakumar, Sadia Mir and Tanya Kane for commenting on various abstracts for my presentations. I discussed my book's findings at the annual conferences of the British Society for the History of Science at the University of St Andrews and the British Society for Middle Eastern Studies at the University of Wales Trinity Saint David. I also had the good fortune to speak at a meeting of the Canadian Society for the History and Philosophy of Science at Brock University and the Three Societies Meeting at the University of Alberta. I also shared the preliminary results at two Liberal Arts International Conferences at Texas A&M University in Qatar. Finally, thanks to Eric Vallet, I was able to present my research at the third edition of the Rendez-vous de l'Histoire du Monde Arabe at the Institut du Monde Arabe in Paris. Outside of conferences, I lectured on this book's topic at Charles University in Prague, the Middle East Institute at the National University of Singapore and Ca' Foscari University of Venice. I again thank Stefano Taglia as well as Linda Matar and Matteo Legrenzi for their invitations to speak at these venues. In Doha, I discussed my research twice at the Qatar Faculty Forum moderated by Ben Reilly at Carnegie Mellon University. I also participated in a research seminar at the Qatar Faculty of Islamic Studies, Hamad Bin Khalifa University.

This book could not have been written without a number of librarians and archivists who provided me with access to sources in Arabic and European languages. Carla Chalhoub, Kaoukab Chebaro (whose first name incidentally means 'planet') and Samar Mikati Kaissi kindly helped me access papers from the Mansur Hanna Jurdak Collection at the American University of Beirut. Ellen Bouton from the National Radio Astronomy Observatory Archives in Charlottesville, Virginia, provided me with a recording of an interview with May

Kaftan-Kassim and further sources on the Iraqi National Astronomical Observatory. Tad Bennicoff from the Smithsonian Institution Archives generously sent me copies of documents relating to Farouk El-Baz and Salah Hamid. James Peters, Anne Anderton and Gwen Riley Jones from the University of Manchester located documents about the Helwan Observatory among the Zdeněk Kopal papers and digitised them for me. Alice Toby-Brant, Claire Hatty and Adam Perkins from the Cambridge University Library generously helped me access correspondence from the Royal Greenwich Observatory Archives. I am also thankful to Amy Andres and Lore Guilmartin for ordering numerous books for me for the library at Virginia Commonwealth University in Qatar. Finally, I am deeply grateful to my interviewees for sharing with me documents from their personal archives. Special thanks go to Emily Piaget Johnson from Boston University for sending me numerous photographs from Farouk El-Baz's collection.

Besides academic support, I was fortunate to receive enormous amounts of care, kindness and encouragement from my family, especially Christian, Claudius, Michael, Sibylle, Peter and Sophia. My parents sparked my interest in the Universe when we were watching the beautiful night sky in Death Valley, California, on a holiday during my childhood. I also thank them for allowing me to watch countless episodes of *Star Trek*, even during times when my marks at school should have demanded more attention. My grandfather, Fritz Determann, who developed telescope software for Carl Zeiss AG, has been a constant source of scientific knowledge. Finally and foremost, I thank my wife, Jeanne Juliana Vaz, for generously reading and editing my work, and for providing wonderful conversations. I proudly acknowledge that she lent me her inspiring dissertation on Cold War scientific and cultural exchange.[1] Above all, I am always thankful for her infinite love.

## STYLE

In writing Arabic names, I have tried to use forms that are recognisable or can be looked up easily in library catalogues, databases and encyclopaedias. Rather than imposing a single transliteration system, my book reflects the diversity of spellings of

Arabic names by Arabs themselves. When authors or journals have a preferred way of spelling their names in Latin characters, I have used that one (for instance, Nidhal Guessoum and *Al-Muktataf*). For Arabic words that have entered the English language, I have used the forms in the *Oxford Dictionary of English.*[2] In my bibliographical references, I have followed the Romanisation standards of the American Library Association and the Library of Congress. I have adhered to the same standard for the transliteration of Arabic words without a common transliteration. In the main text, however, proper names are usually given without diacritics. All amounts given in dollars are US dollars. The citations broadly follow the notes and bibliography format in the *Chicago Manual of Style.*[3] I am very grateful to the developers of Zotero, a tool that has been invaluable for citing and many other research purposes.

# CHAPTER 1

# The Arab Conquest of Space

In February 2013, the First Doha International Astronomy Conference opened at the Qatar National Convention Center. The co-chair Khalid Alsubai from Qatar Foundation (QF) took the podium and declared, 'Astronomy has come home.' After a keynote presentation by Martin Rees, the British Astronomer Royal, Alsubai presented 'Astronomy in Qatar'. He announced QF was planning to establish an astronomy and space centre that would include an observatory and a planetarium. He also hoped that the Qatar Exoplanet Survey, which he headed, would discover two or three new extrasolar planets before the end of the year.[1] A few months after the conference, and partly as a result of its success, Alsubai won a $5 million grant from the Qatar National Research Fund to expand his survey.[2]

The conference and grant marked highpoints for Alsubai, who had spent several years searching for extrasolar planets. In 2010, his team had put Qatar on the map of other worlds – and the map of global astronomical research – through the discovery of Qatar-1b. This was a Jupiter-sized planet orbiting a star 550 light years away. 'We feel proud that one of our citizens, in collaboration with his foreign colleagues was able to discover a new planet which holds the name of our beloved country,' said one Qatari national. 'Historically, Arabs are pioneers in astronomy. So we will not be surprised when one of their successors discover a new planet,' he added.[3] In 2011, the Qatar Exoplanet Survey discovered another Jupiter-sized planet, Qatar-2b, around 500 light years away.[4] Alsubai admitted that his team had faced resistance when seeking to name planets after a country on Earth. However, the astrophysicist was able to convince the International Astronomical Union that he was merely continuing a centuries-old tradition of giving Arabic names to stars.[5] He also claimed that he was, in fact, naming the planets after his survey,[6] which was an established practice. The planet WASP-3b, to whose discovery in 2007 he also contributed, was named after the Wide Angle Search for Planets (WASP), for instance.[7]

The Qatar Exoplanet Survey was not just the project of a particular Gulf state, but part of a broader Middle Eastern space science picture. When Alsubai spoke of astronomy's 'home', he could have meant the region as a whole. The Doha conference's first session, during which he presented 'Astronomy in Qatar', also included an invited presentation by Habib Khosroshahi from the Iranian National Observatory. This presentation was entitled 'Astronomy in the Middle East – cultural heritage and current developments.'

Like Khalid Alsubai, many Arab space scientists act on both a state and regional level. Sheikh Salman Althani, another prominent Qatari astronomer, has supported national as well as wider-ranging Arab space projects. He had already established the Qatar Center for Space Science and Astronomy by 2009.[8] In addition, Sheikh Salman was one of the most important patrons of the Arab Union for Astronomy and Space Sciences with its offices in Jordan. As early as 2008, he promised to fund a £2 million observatory in the Hashemite Kingdom, which he described as 'his second country'.[9] In 2013, he visited the Regional Center for Space Science and Technology Education for Western Asia in Amman. On this occasion, Sheikh Salman promised support for the establishment of a planetarium and telescope in Jordan. Furthermore, he discussed the creation of a database with information on Arab institutions and individuals in space science.[10]

The actions of Khalid Alsubai and Salman Althani are testament to the important role of space science for even a small Arab country such as Qatar in the last decade. However, these initiatives also raise further questions. Is Qatar, metaphorically, trying to conquer the universe, spreading its brand across the galaxy by naming planets after itself? This question may seem strange to some, but perhaps so did Qatar's ambitious plan to host the 2022 FIFA World Cup. Known also for its broadcaster Al Jazeera, the country has been described as a 'small state' engaged in 'big politics'.[11] However, can we expect Qatar or any other Arab country to shape 'big' space politics in the future? How serious and successful has the Arab conquest of space been so far? To answer these questions requires an investigation into the history and politics of modern Arab astronomy and its applications. This book is such an investigation, covering astronauts, observatories, rockets, satellites and other related topics.

Unlike many other histories of Arab science,[12] my book thus focuses on a period much later than what has often been described as the Islamic 'Golden Age'. The latter usually comes before the European Renaissance, which drew heavily on Muslim learning. In contrast, my book concentrates on Arab space exploration since the middle of the 19th century. I am thus covering a period that has often been labelled the *nahḍah*, meaning the 'renaissance' or 'awakening', especially in Arabic literature. This renaissance in turn presupposes a period of 'decline' or 'stagnation' following the 'Golden Age'. Like many other science historians, I do not find vague concepts such as 'Golden Age' or 'decline' helpful for my analysis. I will therefore refrain from labelling modern astronomers such as Alsubai as representatives of an Arab 'renaissance'. Nevertheless, I recognize vague and flexible concepts such as 'Golden Age' as being powerful ideas in themselves. As such, they have the potential to shape modern science just like any other powerful thought.

Rather than seeing Arab space science as a continuation of, or return to, the past Golden Age of learning, I consider it to be fully part of a modern enterprise. The concept of 'modernity' can, of course, be just as vague as that of 'renaissance'. In the following section, I will thus distinguish modern Arab astronomy from its pre-modern form. The differences between science in contemporary Doha and in medieval Damascus lie not just in technologies, but also institutions and ideologies such as the modern state, nationalism and internationalism.

## ARAB WORLD SCIENCE

My book will put modern Arab astronomers in their global as well as national and regional contexts. Transcending the Middle East, initiatives such as the Qatar Exoplanet Survey are also part of world science. They thus differ from traditional Arabic and Islamic learning by belonging to global communities rather than an Islamic empire or civilization. The discoveries of Qatar-1b and -2b were first reported in the *Monthly Notices of the Royal Astronomical Society* and the *Astrophysical Journal*, rather than in an Arabic periodical. The authors of both articles were affiliated with institutions in America and Europe in addition to Qatar.[13] Instead of adopting a specifically Islamic theme, the organisers of the First Doha International Astronomy Conference chose 'Gravitational Microlensing: 101 Years from Theory to Practice.' In their rationale, the organisers did not refer to any Arab or Muslim scientist, but instead commemorated a notebook entry by the German Jew Albert Einstein. This entry postulated that the gravitational field of a foreground star would bend the light of a more distant star.[14] Like Alsubai's survey, Salman Althani's support for astronomy in Jordan also had a global as well as Arab context. The Regional Center for Space Science and Technology Education for Western Asia primarily served Arab countries, although it was also affiliated with the United Nations.[15] Hosted by this centre was the Arab Union for Astronomy and Space Science. The union was in turn a member of both the Arab League and the International Astronomical Union.[16]

The character of Arab space science as being national, regional and global is manifest in the use of both Arabic and European languages. This use goes back to the birth of a distinctly modern form of Arab astronomy in the 19th century. As early as 1858, the Egyptian Mahmoud Effendi, also known as al-Falaki, meaning 'the astronomer', published an article about the Arab calendar in the French *Journal asiatique*.[17] Many of his successors, who were modern Arab astronomers, continued to write in different languages. The *Journal of King Abdulaziz University*, for instance, featured research articles in Arabic and English together with titles and abstracts in the respective other language. New acronyms also arose from combining

Arabic and English. In 2015, Alsubai and his colleagues named a machine-learning method after the tree that formed the logo of Qatar Foundation: Signal Detection using Random-Forest Algorithm (SIDRA).[18] The following year, Alsubai's team at QF presented another algorithm called DOHA, although without explaining whether this was an abbreviation.[19]

While Arabic continued to be used, space science in the region differed from traditional Islamic astronomy in many of its technologies, most notably telescopes, rockets and computers. Although observatories had existed in Arabic countries for more than a thousand years, they were only equipped with telescopes during the 19th century. A 13th-century Syrian manuscript described gunpowder-propelled 'Chinese arrows'.[20] However, the region's first rocket science program only emerged after World War II. To give another example, one of the classic genres of Arabic astronomical writing was the *zīj*. This was a handbook for computing the positions of stars, planets and the Moon. Scholars in Egypt continued to produce handwritten commentaries on these ephemerides well into the 19th century.[21] Their successors, however, increasingly used printed star catalogues and later electronic databases. The research leading to the discovery of Qatar-1b, for instance, relied on the European Southern Observatory's Data Archive and NASA's Astrophysics Data System.[22] Subsequent observations of RR Lyrae variable stars by the Qatar Exoplanet Survey used the SIMBAD and VizieR services provided by the Strasbourg Astronomical Data Center.[23] The names of these services were, however, perhaps their main connection to the *One Thousand and One Nights*. SIMBAD stood for Set of Identifications, Measurements and Bibliography for Astronomical Data.[24]

Because of the myriad modern technologies, it would be wrong to see Arab space science as a mere continuation of traditional Arabic and Islamic astronomy. Consequently, one can indeed be surprised by projects such as the Qatar Exoplanet Survey. Moreover, the causes of the successes and failures of these contemporary projects should be found in modern Arab and global history. For this reason, this book largely refrains from telling the history of traditional Arabic and Islamic astronomy until the 19th century. You can easily find other literature on this topic by authorities such as George Saliba and David King.[25]

When focusing on the period between the 19th and 21st centuries, the rationales of modern states become more apparent in space exploration. Robert Harding showed how a quest for national power, security and development shaped space policy in developing countries.[26] James Clay Moltz described 'national motivations, regional rivalries, and international risks' in a book entitled *Asia's Space Race*.[27] Other scholars such as Ajey Lele and Asif Siddiqi questioned whether this race was more 'hype' and 'rhetoric' than 'reality'.[28] If Asian countries were not in a race, they seemed to be at least in competition with one another. In either case, the achievement of national aims is part of the logic that has been driving modern states to invest in space exploration.

Within the broader field of space science (or space *sciences*), some investments perhaps make more sense than others when using the logic of nation states. Astronautics appears to be a more obvious target for growth than astronomy, for instance. In other words, some governments may have been more easily convinced to fund the development of advanced rockets than telescopes. After all, rockets can be used more obviously for both military and civilian purposes. In English, 'missiles' are generally understood as weapons, but 'rockets' not necessarily so. In Arabic, however, the terms 'missiles' and 'rockets' are often conflated as *ṣawārīkh*, the plural of *ṣārūkh*. Middle Eastern conflicts have indeed partly driven rocket programs in Egypt and Iraq as well as in non-Arab countries such as Iran and Israel. Regionally, arms races have hardly produced peace and stability, of course. However, to an individual government, rockets could still appear to contribute more to national security than astronomical observatories.

With investments in rocket science being more obvious, growth in other areas of space science, especially astronomy, requires more explanation – and more lobbying by practitioners. What could an initiative such as the Qatar Exoplanet Survey have offered to the government that was backing it? Alsubai argued that, 'if you want to build a rich country with rich people, you need to build science and love for science.'[29] Such an economic argument focuses on the very long term, however, and could also be used to support sciences other than astronomy. Alsubai admitted that 'it isn't easy to convince people or the government to finance something like this because society doesn't support astronomy. None of the Arab countries are working on a project like this and Qatar should be proud of its investment.'[30]

Nevertheless, even in Qatar, individuals and institutions were sceptical about the returns on astronomical research. The success of the Qatar Exoplanet Survey is thus still puzzling. 'One of the biggest problems for Qatar is that we lose a lot of young people to business and accountancy studies,' Alsubai said in 2011, 'very few go into research.'[31] Among the various areas of research, Qatar Foundation (QF) was also not immediately convinced about investing in 'globally competitive and mature fields such as astrophysics'. A book on *Arab Science* published by QF in 2009 stated, 'experts believe Qatar and others in the Arab region have the chance to become competitive in newer, largely unexploited fields such as solar and alternative energies, desalination, biotechnology and nanotechnology, and the impact of climate change.'[32] In order to attract funding, Alsubai thus sought to link his research to practical applications. The Cold War space race 'taught us a lot about how to use satellites, which are now integral to life as we know it', he said. 'People don't consider this when they think such exploration is a waste of money.' The astronomer added that the discovery of large amounts of carbon dioxide in Venus's hot atmosphere provided a 'model of Earth's future'. 'Many "green" companies are now making profit from trying to combat global warming,' Alsubai added, 'but we need to remember where this inspiration comes from.'[33]

Alsubai was not the only Arab astronomer who lamented a lack of interest in his discipline. Elsewhere in the region, astronomers were also struggling. In 2013, Nidhal Guessoum, an Algerian professor at the American University of Sharjah, published a four-page comment in the journal *Nature* entitled 'Time for an Arab astronomy renaissance'. He found that most Arab countries were generating fewer than ten papers in the field each year. Only Egypt produced similar numbers as Israel and Turkey – between forty and sixty. Degree programs in astronomy or astrophysics at Arab universities 'can be counted on two hands', he wrote.[34] Guessoum blamed this neglect of astronomy on a 'strongly utilitarian Arab Muslim approach to science'. In the second half of the 20th century, newly-independent Arab states focused on applied sciences such as petrochemical engineering and pharmacy. Astronomy, in comparison, seemed to require 'expensive buildings, equipment and technicians for little tangible return'.[35] This focus on subjects with immediate applications for development makes it even more surprising that the Qatar National Research Fund financed Alsubai through its flagship National Priorities Research Program.

One could, of course, argue that a wealthy state precisely like Qatar could afford planet hunting – and astronomical research in general – as a luxury. Guessoum acknowledged that 'funding should not be a serious constraint' in the Gulf states.[36] The states' distribution of oil wealth among their societies has allowed individuals to develop their passions even without much direct government support. The Kuwaiti merchant Saleh Alaujairy, for instance, established his own observatory in the 1970s, purchasing a three-metre dome and telescope for it in America.[37] Before gaining Qatar Foundation's support, Khalid Alsubai also financed his astronomical studies privately. The Qatar Exoplanet Survey was previously named the Alsubai Project. Moreover, the astronomer even employed a postdoctoral researcher through his Alsubai Establishment for Scientific Studies.[38] Individuals from wealthy Arab countries were also able to support a few other projects elsewhere. Salman Althani supported astronomy not just in Jordan, he also donated three small telescopes to a science centre in Gaza in 2013.[39] Poorer Arab countries such as Mauritania and Yemen still produce science fiction writers.[40] However, compared with Qatar or Kuwait, they do not have the same resources to support actual astronomical careers.

However, financial resources alone cannot explain the success of projects such as the Qatar Exoplanet Survey. Oil was neither sufficient nor necessary for countries to invest in space science and technology. Some oil-poor states in the Middle East, including Israel, were among those most interested in outer space. In 1990, the United Nations proposed the establishment of the Regional Center for Space Science and Technology Education for Western Asia, although only Jordan and Syria then offered to host it. A similar UN-affiliated institution was the Centre Régional Africain des Sciences et Technologies de l'Espace en Langue Français. However, it was set up in Morocco, rather than its wealthier neighbour Algeria, in 1998.[41] Yuval Ne'eman, founder of the Israel Space Agency, even argued that his country had become a space-faring nation because it 'does not have natural resources. Therefore, the essence of its economy, social development and most of all security needs, derive from its only resource, highly qualified man force in Science and Technology.'[42]

In addition to looking at a country's overall resources, it is worth considering individual interests, determination and mobility when explaining Arab space projects. This book therefore investigates biographies of individuals as well as the logic of states. Khalid Alsubai, like other astronomers, became interested in the stars as a child through reading and watching the night sky. Although he was unable to study astronomy at Qatar University and began a career in petroleum engineering instead, he kept his passion for the subject alive. He later went on leave from a well-paid job at Qatar Petroleum in order to study astronomy at Australian and British universities, despite his friends mocking him. 'Why can't you stay on the surface of the Earth?' one of them said, 'you either go deep underground to find oil or to the sky, to the stars!'[43] King Abdullah's personal interests may also have helped Jordan to become a regional center for space science. As a prince, he made a cameo appearance in 'Investigations', an episode from the television show *Star Trek: Voyager*, which first aired in 1996. Fifteen years later, the King Abdullah II Fund for Development became the main investor in the Red Sea Astrarium, a $1.5 billion resort in Aqaba that included a Star Trek-themed space flight adventure.[44]

On the other hand, an individual ruler's *lack* of interest could arguably be equally important for space policy. Muammar Gaddafi, for instance, backed gigantic technological projects on Earth. They included the Great Man-Made River, a huge network of water pipes under the Sahara Desert. However, he was very critical of space exploration. (This is perhaps ironic, as his last name – al-Qadhdhāfī – has the same root as *qadhīfah*, another Arabic word for 'rocket'.) A writer 'at the frontiers of literature',[45] Gaddafi once published a short story entitled 'Suicide of the Astronaut.' In this story, a man travels to the Moon and 'finds nothing'. On returning to Earth, the space explorer is not qualified for basic work in trade or farming. Unable to find a job, he kills himself.[46]

Perhaps partly because Gaddafi did not see much of a future in space, not just Libyan astronomy, but also rocket science remained limited. One initiative was a launch facility in the Sahara Desert that was established between 1980 and 1981. It was constructed and operated briefly by the German company OTRAG, a company that had also worked in Zaire. OTRAG was an acronym for Orbital Transport-und-Raketen Aktiengesellschaft, meaning 'Orbital Transport and Rockets, Incorporated'. With its headquarters in Munich, the corporation designed a small rocket to be used in groups to lift large payloads. OTRAG defended the rocket as being a peaceful, commercial alternative to other launch vehicles. However, by the end of 1981, international pressure had forced the company to halt its Libyan activities.[47]

In order to understand how the interests of individual scientists and rulers can flourish, it is also important to investigate institutional settings. This book will thus move between biographies of individuals and the histories of universities and observatories. Israel's big science, including its space program, was not just the product of an individual's vision, but also of a 'triangle of government agencies, industry and academia'.[48] Among the Arab countries, Egypt and Iraq perhaps came closest to having a military-industrial complex as sophisticated as Israel's during the 20th century.[49] In the 21st century, however, even small Arab countries have had ambitions to produce science and technology on a large scale. When Khalid Alsubai's began his exoplanet survey, Qatar Foundation, for instance, was attempting to establish a close network between the organisation's member institutions. They included the branch campuses of foreign universities, the Qatar Science & Technology Park and the Qatar National Research Fund.

Without a strong military-industrial complex, many Arab countries have restricted themselves to importing rockets and developing them only in minor ways. In other words, the achievements of Libyan space science, for instance, have remained modest not just because of disinterest by Gaddafi personally, but also due to a lack of institution building in his 'state of the masses' (*jamāhīrīyah*). When OTRAG withdrew from Libya, Gaddafi's state had little capacity to continue the German rocketry program on its territory. Libya was still importing Scud-C missiles from the Soviet Union in the late 1980s. Reportedly, the *jamāhīrīyah* was also working on a missile with a range of 750 kilometres. It was named al-Fātiḥ, meaning 'the conqueror'. However, international sanctions affected the Libyan missile program severely and left it underdeveloped compared with the Iraqi one of the 1980s.[50] In the 2000s, Gaddafi gave up his programs to develop weapons of mass destruction in exchange for a re-integration into the international community.[51]

Even when well-funded, institutions can lack dynamism and effectiveness without ideas and ideologies. This book will thus explore the symbolic as well as material resources of Arab space science. As elsewhere, many leaders in the Middle East have employed nationalism and national interests to justify spending on science. Although this book does not focus on Israel, another reference may be useful for understanding Arab space science from a comparative perspective. As early as 1949, David Ben-Gurion, the first prime minister of Israel, told a graduating class of officers: 'The development of Israel, the advancement of its agriculture, industry, seafaring, education and the restoration of the nation – all entail the fostering of science to the limit of our mental and material capabilities.'[52]

By the 2000s, the connection of science with national interests had become explicit even in Qatar, a latecomer in terms of nation-building. In 2009, Tidu Maini, an engineer and executive chairman of the Qatar Science and Technology Park, described his institution as a 'unique experiment in accelerating research in a nation where education and healthcare are the centerpiece of national strategy and intent.' Maini, who had worked as pro-rector of Imperial College London during the early 2000s, even projected his idea of national science back onto the regional past. He added that 'the golden age of Islamic science flourished because of a fundamental recognition that free thought, education and the health of people was vital to a nation's strength and prosperity.'[53]

One of the reasons astronomers such as Khalid Alsubai gained institutional support was perhaps because they connected their research to visions of national development. As part of the Qatar Foundation Annual Research Forum in 2010, Alsubai chaired an Astrophysics and Aerospace Workshop. In a clever and subtle way, he thus attached his research to the aviation field, in which his country already owned a globally competitive company, Qatar Airways. After a presentation entitled 'Aerospace R&D in Qatar – Developing a National Industry,' the workshop agenda included three papers on extrasolar planets.[54] Parallel to his research, Alsubai promised to contribute to education. In 2013, he told the Qatar National Research Fund, 'We will develop the human resources, and we will be able to prove to the public that astronomy is a very good introduction for science in general.' Alsubai added, 'we have a big problem in Qatar – we need to attract university graduates to science.' He therefore launched Fascination Astronomy, a $1 million outreach program in collaboration with Qatar's Supreme Education Council. This program organised school visits, workshops for students and teachers as well as the Qatar Astronomy Olympiad.[55]

While nationalism was part of space projects all over the world, some ideas were specific to Arab and Islamic countries. Although science in general, and space science in particular, were often associated with the future, these ideas were rooted in a common past. Powerful concepts included those of a Golden Age and a renaissance of Arabic and Islamic science. These concepts have been criticised as 'creatures of nostalgia'[56] and products of orientalist scholarship.[57] However, these ideas were important, as they entered the thinking of patrons and administrators of Arab scientific institutions. Moza bint Nasser, chairperson of Qatar Foundation, stated that QF's book *Arab Science* 'commemorates the inauguration of Qatar Science and Technology Park by tracing the proud history of Arab science and invention and its current-day renaissance.' She added, 'The Qatar Science & Technology Park represents a re-awakening of our Arab heritage of invention, inquiry and excellence.' Tidu Maini expressed similar thoughts, 'We are waking to a new dawn made possible by the vision of the Emir and Her Highness and in memory of the golden age of Islamic science.'[58]

Ideas of a past Golden Age and its renaissance have thus clearly shaped contemporary Arab policies in science and technology. Nevertheless, because of the profound differences between modern and traditional astronomy in the region, this book will retain its focus on the time period since the 19th century. With the Qatar Exoplanet merely as a starting point, I will analyse Arab space science as broadly defined. In order to assess successes and failures, my book studies individuals, institutions and ideologies on national, regional and global levels. I ask what kind of connections enabled Alsubai and his peers to carry out their research, even when governments and large sections of society lacked interest. Did nationalism endow certain space projects with enough prestige to offset questionable economic returns? How exactly did the idea of a past Golden Age serve as a resource for bringing about something like an 'Arab astronomy renaissance'? If national and regional resources alone were insufficient, to what extent did Arab space scientists need to search beyond their region for support?

These questions are not only interesting from a historical standpoint, they also have implications for contemporary and future space policy. Should individual Arab countries focus on national capacity building and projects such as the Qatar Exoplanet Survey? Are regional organisations such as the Arab Union for Astronomy and Space Sciences able to overcome inter-state rivalries and provide synergies? Or would it be more beneficial for Arab scientists to collaborate with partners outside their countries and region? This collaboration could take place directly or through organisations such as the International Astronomical Union or the United Nations.

## CHAPTERS IN ARAB SPACE EXPLORATION

This book seeks to answer these questions through a series of case studies focusing on different Arab countries and their regional and global scientific networks. Chapter two will lay out arguments and use of theories in more detail, giving examples from different times and places. As this is a history book, the remaining chapters broadly follow a chronological structure. Chapter three starts with observatories and solar eclipse expeditions during the 19th century and the first half of the 20th. This chapter pays special attention to the international networks of the Syrian Protestant College, which became the American University of Beirut. Chapter four continues with Arab contributions to the Apollo program during the 1950s and 1960s. This part also discusses the legacy of the Apollo–Soyuz Test Project and related science for the development of deserts. Chapter five follows the growth of the Arab satellite industry from the 1960s onwards. This chapter pays special attention to the Arab Satellite Communications Organization (Arabsat), which also contributed to sending the first Arab and Muslim astronaut into space. The final chapter discusses 21st-century Mars missions and the turning of Arab space science into a heritage and source of inspiration for future generations. This concluding part of the book thus links my narrative back to the traditional Islamic astronomy of the 'Golden Age', an important part of Arab *turāth*, meaning 'heritage'. Because each chapter follows certain threads such as the fate of specific institutions, projects and individuals, the different chapters overlap chronologically.

In terms of countries, the rest of the book looks beyond Qatar in order to cover the most important contributors to space science in the Arab region. Chapter two pays special attention to rocket science in Algeria and Iraq. However, as a more theoretical chapter, it also gives examples from many other Arab countries. Furthermore, this part makes a few references to Iran and Israel. It does so not because the populations of these countries include large numbers of Arabs but rather in order to compare and contextualise Arab space science within the broader Middle Eastern political arena. Chapter three focuses on observatories in Lebanon and describes solar eclipse expeditions to Libya. As this chapter will show, transnational networks involving American scientists connected Lebanon and Libya as early as the beginning of the 20th century.

Because of its long engagement with space science, Egypt deserves its own chapter as well as mentions in others. The fourth chapter therefore tells the stories of the Helwan Observatory near Cairo and the international careers of prominent Egyptian planetary scientists Farouk El-Baz and Essam Heggy. Chapter five pays special attention to astronauts and cosmonauts from Saudi Arabia and Syria. They include a son of King Salman and a fighter pilot from Aleppo who became a refugee during the Syrian Civil War. The final chapter mentions various countries again, but puts a particular emphasis on the United Arab Emirates and Morocco. As this chapter will show, the Oukaïmeden Observatory south of Marrakesh contributed to the discovery of ultimately seven planets around the star TRAPPIST-1.

Although this book aims at a broad coverage, it cannot be comprehensive in any way. Instead, I seek to complement other works that focus on more specific aspects of the history of Arab space science. To give some examples, Daniel Stolz analysed astronomy in late Ottoman Egypt in an excellent dissertation at Princeton University.[59] Marwa Elshakry reconstructed the history of the pioneering Arabic scientific periodical *Al-Muktataf*, which popularised modern astronomical knowledge in the region.[60] Owen Sirrs gave a very good account of Egyptian missile programs, but did not rely on many Arabic sources.[61] While I refer to these and many other authors, I do not wish to repeat most of their findings.

Throughout the book, I will pursue the argument that nationalism and cosmopolitanism have equally driven and shaped Arab space science. In the next chapter, I will broadly analyse how they did so. Nationalism and cosmopolitanism – like 'modernity' – are, of course, very vague and multivalent concepts. Even when applied just to the Middle East, they can mean very different things to different people at various times. Nationalism can either refer to territorially defined states such as Qatar, or to an ethnically, culturally or linguistically defined pan-Arab nation. In the former sense, it is often translated into Arabic as *waṭanīyah*, and in the latter as *qawmīyah*. However, there is still considerable ambiguity even in the Arabic. The word *waṭan*, meaning 'homeland', can refer to a nation-state or – in *al-waṭan al-ʿArabī* – to the entire Arab world. Cosmopolitanism evokes images of 19th-century cities such as Alexandria and Cairo under British colonial rule, but also of futuristic, 21st-century Dubai. Nationalism is sometimes associated with ethnic or civic homogeneity and loyalty, whereas cosmopolitanism is linked to diversity and lack of allegiances to specific nation states. Despite the apparent contradictions, however, these phenomena can coexist in the same people, institutions and projects. There are 'cosmopolitan patriots' and 'cosmopolitan nationalism'.[62]

I will refrain from imposing a strict definition of nationalism and cosmopolitanism to the variety of experiences of Arab space scientists. Nevertheless, I emphasise two general tendencies. Nationalism is often concerned with national power, sovereignty and security. Nationalist space projects are therefore primarily aimed at strengthening and protecting individual Arab states or a wider Arab 'homeland' (*waṭan*) or 'nation' (*ummah*). Cosmopolitanism is more concerned with crossing national borders and belonging to diverse communities and global cultures. A cosmopolitan space scientist often works with collaborators from different countries and frequently invokes 'humankind' (*basharīyah*), 'humanity' (*insānīyah*) and the 'world' (*ʿālam*). Playing on a similarity in the verbal roots, he could thus be described in Arabic as *ʿālim ʿālamī*, a 'global scientist'. I am especially interested in instances where nationalist and cosmopolitan tendencies intersect. An example would be a space mission that is launched for nationalist reasons, but ends up producing a vision of the planet without borders.

Cosmopolitanism is perhaps an old-fashioned and inappropriate term in certain contexts. Perhaps for that reason, the literal Arabic translation, *kūsmūbūlītānīyah* or *kūsmūbūlītīyah*, is rarely used compared with words such as *ʿālamī*, meaning 'global', or *ʿawlamah*, meaning 'globalisation'.[63] Throughout this book, I will therefore also use other concepts for phenomena that transcend countries and regions. These concepts include internationalism, international collaboration, transnational networks and global circulations. Nevertheless, the term 'cosmopolitan' often produces more tension with 'national' because it does not contain the word 'national' (unlike trans*national* and inter*national*). Moreover, in a book on space science, 'cosmopolitan' sometimes seems more appropriate than 'global', as the word 'cosmos' can refer to the universe as well as Earth. Astronauts can become more than transnational and global actors, as they not only cross borders, but leave the planet altogether.

# CHAPTER 2

# Nationalism and Cosmopolitanism

In 1997, the Syrian scientist and science fiction writer Taleb Omran first published an Arabic book entitled *Space as Wide as a Dream.*[1] Omran, a professor at Damascus University with a doctorate in differential geometry and astronomy, subsequently republished the book several times. Set in the year 2035, the book describes an 'Arab Science City' on an island in the Arabian Sea. Omran imagined this city to be supervised by 'outstanding Arab scientists, who have come from all corners of the world to contribute to the creation of a modern civilization that returns to their nation its past glory.' The city had advanced laboratories and departments for researching extraterrestrial life and space engineering. The city was 'a dream-like idea that seems impossible to achieve', Taleb Omran wrote, although in the writer's imagination it was realised. The city was born when Arab scientists who had worked in foreign countries built a 'society based on love and cooperation' on the remote island.[2]

Taleb Omran's scientific utopia was also an Arab nationalist one. The 'Arab Science City' was decorated with Kufic calligraphy, one of the oldest forms of the Arabic script. Its members were all Arabs who spoke 'the most eloquent Arabic language' (*al-lughah al-ʿArabīyah al-fuṣḥá*), that is, standard Arabic, without any recognisable 'accent' (*lahjah*). The scientists had escaped the 'foreign monopolies' that controlled the countries where they had previously worked. In order to avoid detection by satellites, they had constructed the city underground. From the beginning of the project, they had also built 'sophisticated jamming stations'. This allowed them to launch in secret a series of rockets called *al-Nafīs*. These spacecraft were probably named after a 13th-century Damascene physician who described the pulmonary circulation of the blood.[3]

## STATES AND STARS

While Taleb Omran's 'Arab Science City' was entirely fictional, the powerful impact of nationalism on Arab space science was not. Nationalism influenced space science in the Arab region by shaping the agendas of states as its main funders. When astronomers such as Nidhal Guessoum complained about the lack of state support for astronomy, they also acknowledged its centrality. One of the factors that made Arab space science 'modern' in comparison with traditional Arabic astronomy was, indeed, the role of states. In pre-modern Islamic lands, individual patrons tended to charge astronomers with specific research and teaching. With the exception of the Maraghah Observatory in 13th-century Iran and unlike many madrasas or hospitals, rulers did not create endowments for observatories. As a result, most observatories did not survive longer than a generation.[4]

Beginning in the 19th century, modern states provided observatories with lasting support independent of the fates of individual rulers. A pertinent example is the Khedivial Observatory, which opened in the Cairo suburb of Abbasia in 1867. Under different names and administrations – but always under the Egyptian state – the observatory long outlived its patron Ismail. It was first under the control of the Egyptian Ministry of War and later, partly, the Ministry of Education.[5] When it moved to Helwan south of Cairo in 1904, it was under the control of the Survey Department of the Ministry of Finance. Renamed the Helwan Observatory in 1914, it remained under the control of the Physical Department of the Ministry of Public Works. Subsequently, it came – as the Royal Observatory – under the control of Fouad I University. After the end of the monarchy, its official name switched back to the Helwan Observatory. By the 1970s, it was part of the Helwan Institute of Astronomy and Geophysics, which itself belonged to the Academy of Scientific Research and Technology.[6]

On the Arabian Peninsula modern state building occurred later, and individual princely patronage for astronomy lasted longer than in Egypt. However, the Gulf region's rulers also started to sponsor astronomers via governmental and semi-governmental institutions rather than personally. The Emir Hamad bin Khalifa and his consort Moza bint Nasser did not fund Khalid Alsubai directly, but through Qatar Foundation and its Qatar National Research Fund. The German co-chair of the First Doha International Astronomy Conference, Martin Dominik from the University of St Andrews, even met Sheikha Moza before Alsubai did, which he did through the Royal Society of London rather than a personal connection.[7] Prince Mugrin bin Abdulaziz, a Saudi governor and later intelligence chief, set up a private observatory, including a charge-coupled device camera and spectrograph. In 1994, he and three professional Saudi astronomers used this equipment to observe the collision between the comet Shoemaker–Levy 9 and Jupiter.[8] However, Mugrin's collaborators, Yaseen Almleaky, Hassan Basurah and Abdulrahman Malawi, were employed by King Abdulaziz University rather than him personally. When the prince donated his observatory in 2009, he did so to the university rather than to any individual client.[9]

As funding became institutionalised, loyalty to a whole regime, rather than an individual patron, became important for Arab astronomers. In Baath-controlled Iraq and Syria, this loyalty was expressed through party membership. Saddam Hussein's government forbade scientists who did not join the Baath Party from holding senior positions. It also denied scholarships to students who were non-members and forced academics to give good marks to students who were. In 1980, the Revolutionary Command Council banned the Islamic Dawa party and sentenced its members to death *in absentia*.[10] Around the same time, May Kaftan-Kassim, who directed the Iraqi National Astronomical Observatory project, felt discredited by certain members of her staff who were Baath Party members. In 1981, she lost her position to Hamid Al-Naimiy, the project's senior astronomer and a party member, and left for America.[11] In 1987, Mohammed Fares would arguably not have become the first Syrian cosmonaut, had he not been a member of the Baath Party for almost fifteen years.[12]

Whether the Arab regimes were monarchical or republican, they demanded loyalty to territorial states. Nationalism and patriotism thus became important parts of space science and other state-sponsored activities. In 1968, an advisory committee headed by the British Astronomer Royal Richard Woolley recommended the construction of a 200-inch telescope near Riyadh. They did so based on 'the attraction it would offer to foreign astronomers, and the prestige which it would confer on Saudi science.'[13] Although this particular telescope remained unrealised, the Saudi Arabian National Center for Science and Technology later undertook a National Observatory Project.[14] Other governments in the region also supported institutions that were explicitly national. In 1973, President Ahmed Hassan al-Bakr ordered the construction of the Iraqi National Astronomical Observatory, which evolved into a project consisting of three telescopes and a total cost of $160 million.[15] Egypt's Helwan Institute of Astronomy and Geophysics was renamed the National Research Institute of Astronomy and Geophysics in 1986.[16]

Nationalism not only left its mark on observatories, it also supported the education of professional astronomers. During the 1920s, the formally independent Egyptian government insisted that any new astronomer appointed by the Helwan Observatory should be Egyptian rather than English. The government therefore sent Reda Madwar, an engineer at the Ministry of Public Works, to study for a doctorate in astronomy at the University of Edinburgh. After his graduation in 1926, he became the resident astronomer and, in 1934, director of the Helwan Observatory. In 1936, he was also appointed to a professorship at the Egyptian University,[17] where he taught the next generation of indigenous astronomers. Several decades later, Egyptians affiliated with the Helwan Observatory and Cairo University set up the first Saudi astronomy department at King Abdulaziz University (KAU). At the same time, KAU sent Saudi students to Britain in order to earn their doctorates and replace their former teachers as part of national capacity building. In short intervals between 1989 and 1991, Yaseen Almleaky and Hassan Basurah completed their doctorates at the University of Glasgow, as did Abdulrahman Malawi at Manchester University.[18]

As well as observatories and university departments, the governments in the region supported rocket science for national prestige and security reasons. The rocket program in Algeria between the 1940s and 1960s was under the control of the French military, although it depended on a lot of German and American help.[19] Several versions of the Véronique sounding rocket were launched from a Centre interarmées d'essais d'engins spéciaux, meaning 'Joint military test centre for special machines'. This centre was located in Colomb-Béchar and Hammaguir in the Sahara Desert close to the Moroccan border. The first French satellite, launched from Hammaguir in 1965, was named Astérix after the comic book character.[20] This was appropriate, as the name alluded to *astérisque*, meaning 'little star'. (Although the name was perhaps ironic, as the Gaulish warrior's resistance against Roman occupation could be seen as mirroring the struggle of the Algerian National Liberation Front against the French.)

Not just the French, but also many Arab governments were attracted to rocket science, in particular because of its military potential and its association with technological progress. One of the first was the Egyptian government, which became interested in rockets after the Arab–Israeli War of 1948.[21] In January 1952, the Egyptian magazine *al-Thaqāfah*, meaning 'culture', published an article entitled 'The German Rocket', that is, the V-2. The article argued that rockets 'have a great future', imagining their use not only in wars, but also for atmospheric science and the transportation of air mail across continents in a 'matter of minutes'.[22] After another war against Israel in 1956, Egyptian interest in this technology intensified. In a 1957 speech, Egyptian President Gamal Abdel Nasser said, 'We must restore the armed forces in light of our experience in the Suez War.' He stressed the importance of planes and rockets in particular. However, he stated that 'the issue is not that we are able to make rockets or planes. What is important is that these areas are the technology of the future, and we must allow the Egyptians to recognize that.'[23]

Under Gamal Abdel Nasser, the Egyptian missile industry was thus created, but not without significant foreign help. A complex at Helwan and a plant in Ain Shams, named Factory 333, were established to support the production of missiles. In 1962, Nasser attended the test launch of two single-stage rockets named al-Ẓāfir, meaning 'the victor', and al-Qāhir, meaning 'the conqueror'. In 1963, the mockup of a two-stage rocket called al-Rā'id, meaning 'the pioneer', was displayed at a military parade. This rocket program largely relied on German engineers who had worked on the V-2 during World War II.[24] For this reason, the program suffered when Israel targeted these scientists with assassination and intimidation, and put pressure on the West German government during the 1960s. In the mid-1960s, the Federal Republic of Germany began a counter-proliferation strategy, seeking to bring its scientists home.[25] By 1967, the Egyptian missile program had declined and played hardly any role in the Six-Day War against Israel.[26]

Unlike in Egypt, the rocket program in Lebanon began as a private and entirely civilian venture under the aegis of the Haigazian College Rocket Society in 1960. However, the military soon gained oversight there as well. Initially, the society shot 'tiny baby rockets' less than half a metre long from a mountain near Beirut. Partly because of security concerns, Lieutenant Youssef Wehbe, a ballistics expert, told the society in 1961 to use an artillery range. Under Wehbe's supervision, the Haigazian College Rocket Society developed two-stage and three-stage rockets. The program also became a source of national pride, with the society being renamed the Lebanese Rocket Society and its rockets 'Cedar' – after the tree that is Lebanon's national emblem. In 1964, an artist's image of a Cedar rocket leaving Earth was the decoration on two stamps commemorating the twenty-first anniversary of Lebanese independence. A final rocket, the Cedar 10, flew under military auspices in 1967, the year of the Six-Day War. Charles de Gaulle had reportedly pressured the Lebanese president to shut down the project for geopolitical reasons.[27]

After the Six-Day War, Egypt rebuilt its armed forces. Convinced of the importance of ballistic missiles in a future conflict, it imported a lot of technology from the Soviet Union. Other Arab countries did the same such as Libya, Syria and especially Iraq. In 1969, the Iraqi army began systematically recruiting soldiers and officers into a special branch for guided missiles. For security reasons, the volunteers had to be Iraqi nationals with an Iraqi father and an

Iraqi or Arab mother.[28] In 1972, Iraqi officials signed contracts for the procurement of two Soviet surface-to-surface missile systems, the 9K52 Luna-M, meaning 'Moon', and the R-17. Referred to by the North Atlantic Treaty Organisation (NATO) as FROG-7 and Scud-B, these missiles had ranges of up to 70 kilometres and 300 kilometres, respectively. In the 1970s and 1980s, Iraq procured 819 Scud-Bs, many of which were launched against Iranian cities.[29]

Between its war against Iran and its invasion of Kuwait in 1990, Iraq briefly pursued a civilian rocket program as well. In 1988, the republic established a Space Research Center under the control of its Scientific Research Council. A law signed by Saddam Hussein stated that this centre aimed to 'keep up with the global scientific developments in this vital field and to utilise them in the service of the country.'[30] Shooting rockets from a base in al-Anbar to the west of Baghdad, the centre was soon engaged in a 'small-scale space race' with Israel. Also in 1988, the Jewish state launched its first satellite, *Ofeq-1*, meaning 'Horizon'. The following year, Iraq tested its three-stage launch system al-ʿĀbid, meaning 'the worshipper'. The first stage consisted of five Scud-Bs grouped together. Acting as a test satellite, the third stage completed six orbits before burning up on re-entry.[31] On that occasion, Hamid Al-Naimiy declared in the magazine *Āfāq ʿArabīyah*, meaning 'Arab Horizon': 'Today, Iraq has entered the conflict arena of space technologies.'[32]

Arab security and strategic interests in space have often been combined with economic ones. Al-Naimiy added in his article that Iraq was 'conscious of the importance of space research and the peaceful use of its technologies for the Iraqi and Arab man's prosperity in his daily life.'[33] In 1990, the astronomer published another article on Iraqi space technology in *Āfāq ʿArabīyah* under the title 'Deterrence of the Zionist Project.' He stated that there was 'no longer any doubt that the rockets, satellites and spacecraft of all kinds and purposes are the wealth of the modern world, providing services and information.' Al-Naimiy thus argued that 'the conquest of space and the use of its science and technologies are no longer just a scientific luxury and a thrill for the masses.' Instead, they 'have become an economic and security imperative that cannot be ignored by any civilised country that aims to achieve comprehensive development and the strength to guarantee its sovereignty and dignity.'[34]

While generally supportive of space science in the region, nationalism as warmongering was also obstructive or destructive. The World Wars drew astronomers from different countries into military service or hindered their international movement. Likewise, the Algerian War started by the National Liberation Front in 1954 resulted in a setback for space science. By the time the war finished in 1962, most French astronomers at the Algiers Observatory had left the country. In 1967, the French military evacuated Hammaguir, launching subsequent rockets from Guyana and tracking satellites from Gran Canaria instead.[35] It took an earthquake in 1980 for the independent Algerian state to invest again in the Algiers Observatory, establishing – through the Ministry of Interior – a Centre National d'Astronomie, Astrophysique et Géophysique. In 1985, this institution was renamed the Centre de Recherche en Astronomie, Astrophysique et Géophysique (CRAAG).[36] Similarly, in Egypt, the Six-Day War in 1967 was a setback for the Egyptian rocket program, which had begun with German help.[37] A planned scientific satellite named 'Star' was never launched.[38] Only about three decades later, in 1998, an Ariane rocket launched the first Egyptian satellite, *Nilesat 101*, from French Guyana.[39]

Similar to the North African cases, Iraq's wars were destructive for space science in the entire Gulf region. During the 1980s, the Iraqi government expelled nationals of Iranian origin[40] and recruited many scientists, including young astronomers, into the war effort. In Iran too, the war and international isolation, though beneficial for the growth of military and aviation industries, delayed the establishment of a national space agency.[41] In 1985, before the Iraqi National Astronomical Observatory became operational, Iranian forces damaged it, causing the German contractors to withdraw from the site.[42] By the time of the Kuwait war, during which the observatory was hit again, the Scientific Research Council had been liquidated.[43] The base in al-Anbar still continued its rocket program for a brief period. However, the centre was damaged during the war in 1991, thus ending its space race with Israel.[44] On the other side of this conflict, Kuwait's Alaujairy Observatory suffered from destruction and looting.[45] Atmospheric pollution from the Kuwaiti oil fires also delayed the site selection program for the Saudi National Observatory Project.[46]

The end of the war in 1991 brought no significant recovery of space science in Iraq, as sanctions crippled civilian and military programs. Many Iraqis were cut off from the international scientific community. Subscriptions to international journals were halted, travel to conferences abroad almost stopped and even scientists who were Baath Party members left their country during the 1990s.[47] Under the terms of the sanctions, Iraq was limited to manufacturing rockets with a reach of less than 150 kilometres. Moreover, the United Nations Monitoring, Verification and Inspection Commission (UNMOVIC) destroyed many of Iraq's remaining missiles. However, the country attempted to develop a number of shorter-range missiles known by names such as Abābīl and al-Ṣamūd.[48] These names signalled defiance and resistance and, like al-ʿĀbid, had religious associations. *Al-ṣamūd* means 'steadfastness', and a related word, *al-ṣamad*, meaning 'the everlasting', is an epithet of God in the Qur'an. *Abābīl* refers to a mythological race of birds that appear in the Qur'an. The birds protected Mecca against an invasion with war elephants from Yemen in 570, the year the Prophet Muhammad was born. Al-Ṣamūd and Abābīl missiles were produced in an underground factory near Baghdad. This factory was named al-Maʾmūn,[49] perhaps after an Abbasid caliph and founder of an observatory.

In 2003, the American-led invasion of Iraq largely brought to an end what had remained of Saddam Hussein's space programs. The US military put Amir Al Saadi, the president's leading scientific adviser, as one of its most-wanted members of the Iraqi government on playing cards for soldiers. A British-educated physicist, Al Saadi had overseen missile development and production, and later liaised with the United Nations inspectors. Within two weeks of the invasion, he surrendered to US forces in Baghdad.[50] He was released in 2005.[51] In the meantime, however, the Coalition Provisional Authority had barred many former members of the Baath Party from employment in the public sector. Among those affected were many scientists and engineers in Iraq's military industry.[52] Hussam Muhammad Amin, a former director of the al-ʿĀbid project, complained that 'the barbaric

American occupation' had not only 'destroyed' Iraq's infrastructure, it had also forced scientists and engineers to work in 'humble fields far from their specialty and long experience in order to secure a decent living for themselves and their families.'[53]

Nevertheless, even many citizens from a 'Republic of Fear'[54] such as Saddam Hussein's Iraq did not depend on one state alone in their scientific careers. Similar to the German rocket scientists and engineers after World War II, many Iraqis found opportunities elsewhere. After the liquidation of the Scientific Research Council, Hamid Al-Naimiy, for instance, became chair of the physics department at the University of Baghdad in 1990. Four years later, he moved to Jordan to become the first dean of the Institute of Astronomy and Space Sciences at Al al-Bayt University. Amir Al Saadi was also lucky. After his release from American custody in 2005, he became an advisor to Qatar Foundation and contributed as project manager to the establishment of the Qatar National Research Fund (QNRF). The first executive director of this fund was Abdul Sattar Al-Taie, another scientific administrator who had worked in Saddam Hussein's Iraq.[55] Ultimately, the work by these two Iraqis contributed to the QNRF's support of the Qatar Exoplanet Survey.

The rarity of their expertise and skillsets certainly helped scientific administrators earn new positions in the region. However, arguably just as important for the exodus of Iraqi scientists was an Arab supra-state identity. As former Baath Party members, men like Al-Naimiy not only spoke Arabic, they mastered the rhetoric of Arab nationalism. While still in Iraq, Al-Naimiy served as secretary general of the Arab Union of Physics and Mathematics between 1985 and 1990.[56] In Jordan, he co-founded the Arab Union for Astronomy and Space Sciences in 1998. He first served as its vice-president and, from 2005, as its president. In the name of this union, he proposed the establishment of a space agency to the Arab League.[57]

## SPACE ARABISM

An Arab space agency was especially attractive for scientists such as Hamid Al-Naimiy who had left their countries, or who otherwise lacked support from their own governments. In addition, this type of organisation would be able to gather resources for more ambitious pursuits, especially in big science. By the 1970s, Egypt was already seeking to develop ballistic missiles through the Arab Organization for Industrialization. Saudi Arabia, Qatar and the United Arab Emirates pledged an initial capital of more than $1 billion for this organisation.[58] According to Al-Naimiy's vision, expressed at a workshop in 2000, an Arab space agency, like its European counterpart, would be useful for 'large astronomical and space science research projects' that were 'too expensive for a single country to support on its own'.[59] Al-Naimiy considered the Arab Union for Astronomy and Space Sciences a 'nucleus' for such an agency and sought 'support' from states in the region.[60]

Arabism in space was not just the vision of astronomers, but also of science fiction writers. In a short story entitled 'Love in the Twenty-first Century', the Egyptian author Raouf Wasfi imagined the launch of the rocket *al-Waḥdah*, meaning 'Unity', from Luxor in 2020. Adorned with 'the flags of all Arab countries', the rocket carried the 'first two Arab astronauts into space'. These two, a young man and woman, had been chosen from thousands of volunteers to 'serve the Arab future in the coming centuries'.[61] Similarly, Taleb Omran imagined regional synergies through an 'Arab Space Research Academy' and an 'Arab Scientific Center for the Study of Stars and Planets' – in addition to the 'Arab Science City' in his *Space as Wide as a Dream*.[62]

An Arab space agency, like the ideal of Arab unity, remained by and large an aspiration. By 1993, the Arab Organization for Industrialization had lost its shareholders from the Gulf and was fully owned by the Egyptian state.[63] Funding for the Arab Union for Astronomy and Space Sciences (AUASS) itself initially came mostly from the Jordanian government and a few individual patrons, including Sheikh Salman Althani. The union's annual income remained at around $35,000 during the early 2000s, enough for less than a quarter of its planned activities. In 2002, Al-Naimiy complained about the 'weakness of financial support' and urged members to put 'pressure on their governments to pay the annual subscription for each state.'[64] The AUASS succeeded, however, in turning Jordan's Regional Center for Space Science and Technology

Education for Western Asia into a de facto Arab institution. The participants in the center comprised only members of the Arab League, including African states such as Libya, Somalia and Sudan. Israel and Turkey were excluded despite their location in 'Western Asia' and Jordan having diplomatic relations with both.[65]

Even in the absence of a common space agency, Arab nationalism had the potential to support astronomy in the region. Commitment to a shared language, culture and identity facilitated academic mobility and collaboration throughout the region. Beginning in the 1970s, many Egyptian astronomers, for instance, left for better paying jobs in Algeria and Saudi Arabia. At the time, this movement was sometimes described as a 'brain drain'.[66] In 1986, two Egyptian astronomers noted that the departure of many of their colleagues 'to work in other Arab countries has caused remaining staff in the Helwan Observatory to confine themselves in two fields, extragalactic research and variable stars.'[67] However, the migration of Egyptian scientists also generated collaborations and joint publications between the Helwan Observatory and King Abdulaziz University in Jeddah, for instance.[68]

Although Arab unity on Earth and in space was elusive, Arabism provided people in the region with important symbolic resources, including memories, dreams and promises.[69] 'Islamic astronomy enjoyed a golden age from the ninth to the sixteenth century,' Nidhal Guessoum began in his *Nature* article in 2013. He then urged Arab states to build new observatories. 'A few medium-sized telescopes (one- to two-metres in mirror diameter) costing a few tens of millions of dollars would allow Arab astronomers to join front-line research,' he promised. The Algerian astronomer ended his article by asserting that 'astronomy has a natural place high in the landscape of Arab Islamic culture. It must be brought back.'[70]

Khalid Alsubai's opinion about the Qatar Exoplanet Survey was similar. The nostalgic memory of medieval Arabic astronomy was thus an important part of space Arabism, as envisioned by scientists across the region. 'People may not expect such a tiny country to discover a planet, but we can. Did you know that Arabs have a huge heritage in cataloguing the stars?' Alsubai asked an interviewer following the discovery of Qatar-1b. 'All the bright stars' names are

Arabic. Sure, our role in astronomy and space has been dormant for centuries. But I think our role is coming back. We will put our names between the stars again.'[71] Not just scientists, but also science fiction writers often drew on figures from the medieval past when imagining an Arab future in space. The Syrian Taleb Omran, for instance, named a spaceship *Ibn Firnās* 'after the Arab scientist ʿAbbas ibn Firnas, who attempted to fly for the first time in history'.[72] Similarly, the Egyptian Raouf Wasfi named a spacecraft *al-Idrīsī* after 'the famous Arab geographer, who lived in the 12th century'.[73]

Space scientists also used a common Arab heritage to promote projects that were confined to existing states and to include these states in a transregional astronomical community. The Riyadh telescope committee, headed by Richard Woolley, noted that 'Arabs were the real custodians of astronomical science for several centuries, so that it is appropriate that they should endeavour once again to promote this science.'[74] In Saudi Arabia's application for membership in the International Astronomical Union during the 1980s, the kingdom's representative emphasised 'the past contribution of Arab astronomers and rich cultural heritage in Astronomy'.[75] Around the same time, Hamid Al-Naimiy introduced the Iraqi National Astronomical Observatory as part of a 'cultural, scientific, and technical renaissance' in his country. At a meeting of the International Astronomical Union in Japan in 1984, he reminded his audience that 'the world received important gifts from the Arabs, particularly from Mesopotamia.' They included the astronomical observatory, 'an invention of the Arabs' and one of 'their permanent contributions to human culture'.[76] Drawing on this collective memory, Al-Naimiy was keen to repair the observatory after Iranian forces had damaged it. In his 1990 article in *Āfāq ʿArabīyah*, he hoped that this would restore the 'heritage of the glorious Arab nation, which had established the foundations of these sciences and whose marks are still clear in the various foreign and Arabic sources.'[77]

Despite conflicts between Arab states and Iran, Arab nationalists such as Al-Naimiy had a broad conception of Arabic civilisation that included astronomical contributions by Iranians. The observatory at the Institute of Astronomy and Space Sciences at Al al-Bayt University in Jordan was named Maraghah after the medieval observatory in

Iran. Al-Naimiy, the dean of the institute, did not find it problematic that the patron of the original Maraghah observatory was the Mongol ruler Hulagu, who had led the destruction of Baghdad in 1258. Nor did Al-Naimiy avoid a potential comparison with Nasir al-Din al-Tusi, Hulagu's Iranian Shiite astronomer.[78] In 2009, Al-Naimiy gave a presentation on 'the role of astronomy and space sciences in Arab societies and cultures' at another IAU symposium in Paris. In this presentation, he described the Iraqi National Astronomical Observatory as also being a successor of the medieval observatories in Baghdad, Samarra, Maraghah and Samarkand.[79]

What helped Al-Naimiy overcome the divides between Arabs and Iranians was that the Arabic civilisation whose memory he and others employed was also an Islamic one. Al-Naimiy described Nasir al-Din al-Tusi as a 'Muslim astronomer' and mentioned that 'many Islamic astronomers' had used the original Maraghah Observatory. The Institute of Astronomy and Space Sciences at Al al-Bayt University included a 'Unit for Reviving the Astronomical Heritage of Arabs and Muslims.' One of the objectives of the Arab Union for Astronomy and Space Sciences was 'researching the contribution of the Arab Islamic civilization in the realm of astronomy and related sciences, as well as the roles of various other ancient civilizations in the region.'[80] In 2013, Mohamed Fathy Saoud, president of Qatar Foundation, described the First Doha International Astronomy Conference as a 'very significant event especially because the study of astronomy is part of Arab Islamic heritage'.[81]

## MODERN ISLAMIC ASTRONOMY

Islam was not just a part of regional heritage, but also a framework for some contemporary research. Arab space science thus included modern Islamic astronomy, and space Arabism was not entirely secular. One of the first activities of the Arab Union for Astronomy and Space Sciences was the Islamic Crescents Observation Project, whose purpose was 'predicting the visibility of the young crescent moon which fixes the beginning of the Islamic Holy Lunar months.'[82] In 2006, the union organised the Fourth Islamic Astronomical Conference with the theme 'astronomical applications in Islamic law' at Al al-Bayt University. The conference focused again on determinations of crescent visibility for the calendar as well as on calculations of prayer times and directions. An additional concentration was 'scientific and cosmic miracles' (*al-iʿjāz al-ʿilmī wa-al-kawnī*) in the Qur'an and Sunna.[83]

Some leading professional astronomers were very interested in reading science in scripture. In 2000, Al-Naimiy published a book entitled *The Universe and its Secrets in the Verses of the Qur'an*.[84] He translated verbal information from the Qur'an into numbers. For instance, he gave the words *kull shayʾ*, meaning 'everything', the value of 360. On that basis, he calculated the distance between Earth and God's throne, which is above the seventh heaven, to be $7.665 \times 10^{12}$ kilometres.[85] The Egyptian solar physicist Mosalam Shaltout engaged in similar pursuits. He was perhaps from the same extended family as Mahmoud Shaltout, a grand imam of al-Azhar under Gamal Abdel Nasser. Mosalam studied at Cairo University, however, and spent most of his career at the Helwan Observatory. Like Al-Naimiy, he served as vice-president at the Arab Union for Astronomy and Space Sciences and participated in the Fourth Islamic Astronomical Conference. He published research papers in English in journals such as *Astrophysics and Space Science*. In addition, he contributed in Arabic to the Saudi magazine *al-Iʿjāz al-ʿIlmī*, meaning 'Scientific Miracle'.[86] He believed that the Qur'an is 'God's eternal miracle', 'does not change', and – unlike the Bible and the Torah – 'does not contradict modern science at all'.[87]

Dual interests in astronomy and Qur'anic studies made some researchers both scientists and religious scholars. Indeed, the Arabic word *ʿālim* (plural *ʿulamāʾ*) is often applied to both types. (The term *ʿilmīyūn* for 'scientists' – in analogy to *scientifiques* in French – is less common.) When speaking of 'heaven' and the 'sky', many Arabic speakers would also use one word, namely *samāʾ*. Another example of a man of modern science and religion is one of Shaltout's teachers, Abdel Hamid Samaha. He worked at the Helwan Observatory and taught astronomy to law students at al-Azhar. In 1937, he gave a lecture on 'cosmological ideas in Al Quran' at Lund Observatory in Sweden. He translated passages from the Qur'an, arguing that they referred to the 'sun's motion', an 'absolute time unit', 'stellar evolution' and the 'existence of life' in outer space. He concluded that these Qur'anic ideas 'in the light of modern science have proved to be perfectly sound and correct'. Lund Observatory published Samaha's lecture as an 'interesting statement by a Mohamedan man of science' while refraining from criticism for 'evident reasons'.[88]

In line with dominant Islamic and modern scientific teachings, many Arab astronomers were hostile towards astrology. Whereas the pretext for the establishment of the original Maraghah Observatory was the quest for better astrological results,[89] many modern astronomers sought to put a maximum possible distance between their practice and astrology. In 2002, the Arab Union for Astronomy and Space Sciences recommended the 'fight against astrology' to its members and to the International Astronomical Union.[90] In 2014, a journalist with the Qatari newspaper *al-Waṭan* asked Sheikh Salman Althani about the relationship between 'astrology' (*ʿilm al-tanjīm*) and 'astronomy' (*ʿilm al-falak*). Sheikh Salman denied that astrology was a 'science' (*ʿilm*), adding that it had grown out of 'ancient man's lack of knowledge of the sky'.[91]

Not all of Arab space science was Islamic, however, as it included prominent Christian astronomers such as the Egyptian Joseph Mikhail, although support for Islam benefitted astronomical research in general. Governments financed observatories such as the one in Helwan, not just because they produced astronomical research, but because they calculated prayer times and dates for the Islamic calendar. In the 1970s, the Saudi government invited Khairy Aly from the Helwan Observatory to set up an Islamic Observatory as part of the Muslim World League. His survey of the Hejazi mountains for a suitable site subsequently benefitted the astronomy department at King Abdulaziz University, whose establishment he also led.[92]

Although religious interests supported Arab space science in general, they were limiting in some instances. In 1968, the Riyadh telescope committee reported on potential difficulties in attracting Western astronomers. It noted that as the Saudis 'are very strict Muslims the social life in Riyadh may appear very restricted to expatriates'.[93] Some Muslim researchers also faced religious sensitivities in the kingdom. One astronomer at King Abdulaziz University complained about the general 'lack' of astronomical teaching. He blamed it on Saudi Arabia's 'general culture' during the 1980s and 1990s, which considered astronomy 'a form of astrology or against religion.'[94]

The Islamic lunar calendar was considered especially sacrosanct and the subject of tensions between some scientists and traditionally-minded Islamic scholars. In the early 21st century, Hisham Abad, a physicist at Riyadh's Alfaisal University, sought to reform it on a lunisolar basis. He claimed support from some religious scholars, but was afraid to publish his research inside Saudi Arabia. 'The Salafis, as you know, have no mercy' (*lā yarḥamū*), he told me in 2016. (He thus played on the concept of *raḥmah*, meaning 'forgiveness', which is often seen as central to Islam.) Abad tried to give a public lecture on, or at least lead a 'discussion', of his research at the King Faisal Center for Research and Islamic Studies, which belonged to the same foundation as Alfaisal University. Although he initially received a positive response, he claimed that the center had refused to host him, because his research was 'outside the Salafi line'.[95]

Gender segregation as promoted by the Wahhabi and broader Salafi tradition also created challenges for amateur astronomy. The Qatar Center for Space Science and Astronomy organised an annual 'astronomical camp' in the desert during the 2010s, but only for men. Rather than providing hands-on training on telescopes, Salman Althani preferred to send women 'training materials via email'. When available, he would use female trainers in order to 'give women privacy'. Conscious of religious sensitivities, Sheikh Salman worked directly with the Qatari Ministry of Endowments and Islamic Affairs in organising astronomical lectures. This way, the lectures were meant to spread not just scientific awareness, but also 'moderation'. They intended to reject both 'religious extremism' and 'scientific extremism'.[96]

Not only in Wahhabi-dominated Saudi Arabia and Qatar, but also in postcolonial and Salafi-influenced Algeria, were astronomical initiatives limited by concerns about Islam and a polarisation between the natural sciences and religion. In the 1980s, the newly established Emir Abdelkader University of Islamic Sciences in Constantine introduced a compulsory course in astronomy for all of its students. It was intended to give future religious and legal scholars enough background to use astronomy in Islamic jurisprudence. Although many students opposed the course as a form of 'Western cultural invasion', it continued for several years. During this period, Nouredine Zettili, a physicist and the university's vice-president, planned the establishment of an institute of astronomical sciences and a planetarium. These plans, however, attracted insurmountable opposition from the students and other universities with a focus on the natural sciences. Although the plans had received ministerial and budgetary approval, a wide-reaching strike resulted in the removal of '(Western) scientific materials' from the curricula and the dismissal of all the administrators supportive of them.[97]

The civil war that broke out between the Algerian military and Islamic groups in 1991 was similarly destructive for astronomy, as were the conflicts in Iraq during the same period. After losing his position as vice-president in 1989, Zettili joined Saudi Arabia's King Fahd University of Petroleum and Minerals (KFUPM). In 1990, he returned to Algeria as president of the University of Blida, where he set up a graduate program in astrophysics. One of his faculty members was Nidhal Guessoum. He had just returned to Algeria after the completion of his doctorate at the University of California, San Diego, and carrying out research at NASA. However, the Algerian graduate program soon ended after the arrest, death or migration of most of the instructors.[98] Zettili moved back to KFUPM as an assistant professor in 1992. Nidhal Guessoum, who left Blida for Kuwait in 1994, wrote: 'Astronomy, like the rest of Algerian science, culture, and education, is in total disarray.'[99]

Marked by the conflicts in Algeria, Nidhal Guessoum continued his struggle over authority with religious and legal scholars. In an article in 2001, he argued that jurisconsults 'base their decisions on simple ideas that date back to the early Islamic era and ignore the huge development that has taken place in astronomy.' The astronomer complained that to determine the start of lunar months, Arab officials were accepting 'doubtful testimony of a layman on the observation of the crescent'. At the same time, these officials 'reject the calculations and explanations of professional astronomers'.[100]

Guessoum also expressed embarrassment over the geocentric views held by some religious scholars. In a 2015 interview, Guessoum was asked about the Saudi preacher Bandar al-Khaybari. On a visit to the UAE, this preacher had responded to a student's question that Earth was 'fixed'. Guessoum said that, 'the world's newspapers, from America to Australia, indeed mock this rejection.' The astrophysicist continued, 'with the fierce extremism that we witness in some Muslim countries, the world asks, "Are the Muslims still in the dark ages?"' The Algerian admitted that there were 'defects in the educational system' in 'some places of the Islamic world'. He thus called for a 'review of the education given to Muslim youth today, otherwise the future will be very dark.'[101]

Although Arab space science had a challenging relationship with Islamic scholarship, it was not entirely dependent on the fate of Muslim-majority countries and their education systems. Many scientists in Algeria and elsewhere in the Arab region had opportunities to study or work in other parts of the world. After his bachelor's degree at the University of Algiers, Nouredine Zettili, for instance, earned his master's degree and doctorate at the Massachusetts Institute of Technology (MIT). This connection arguably helped him become a visiting scientist at MIT in 1998, after which he moved to Jacksonville State University in Alabama to take up a faculty appointment. He remained a traveller rather than an immigrant. In 2013, he became an acting dean of the College of Science and General Studies at Alfaisal University in Riyadh.

Zettili's ability to study and work in countries as different as Algeria, Saudi Arabia and America also raises the question as to what extent he and other Arab space scientists were cosmopolitans or 'citizens of the world' – while still feeling a sense of belonging to a specific state, region and religion. My book contends that this cosmopolitanism, as related to transnational exchange and international cooperation, was – alongside nationalism – another major factor in the development of Arab space science.

## CROSSING THE COSMOS

Cosmopolitanism, like Arab unity, was, of course, an ideal – or ideology – more than reality. Narrowly defined, world citizenship without a world state or world government remained an aspiration.

Nevertheless, the basic idea had proven attractive for theorists since the ancient Greeks, who coined the term. In the 20th century, nationalism contributed to the destruction of conviviality and the cleansing of cosmopolitan cities in the eastern Mediterranean.[102] Similar to nostalgia for a past Golden Age of Arabic science, sadness over a lost Ottoman or European colonial cosmopolitanism marked a lot of scholarship on the Middle East. The same scholarship often focused on elites,[103] and I admit that my book with its focus on scientists does so too.

Nevertheless, cosmopolitan ideas survived the departure of many Europeans from cities, such as Alexandria and Cairo. Rockets, despite their destructive potential, partly inspired utopian thinking after World War II. The 1952 Egyptian magazine article about 'The German Rocket' is a pertinent example. It ended on the following note: 'We hope that scientific progress is directed towards the service of humanity', so that 'our planet becomes a paradise of God after it had become the graveyard of mankind.'[104] During the space age, literary and cinematic visions of leaving Earth further promoted the planet's unity in the Egyptian popular imagination. In 1958, the Egyptian writer Tawfiq al-Hakim first published his play *Voyage to Tomorrow*. The main protagonists, two prisoners who have been sentenced to death, are travelling on a rocket to distant planets with little chance of returning. One of them says, 'we travel without passport, without nationality [*jinsīyah*].' The prisoner then tries to convince his fellow traveller that they retain their 'Earth nationality' (*al-jinsīyah al-Arḍīyah*) or their 'Adamic' (*al-Ādamīyah*) nationality in their chests. He argues that 'Earth is our mother.' She has given qualities preserved in 'our souls'.[105]

In his play, al-Hakim suggests that the astronauts keep their 'Earth nationality', while their 'country nationalities' (*al-jinsīyāt al-dawlīyah*) have been cancelled.[106] However, outside utopias, cosmopolitanism and nationalism, or the commitment to global and national communities and interests, could well co-exist. This is especially the case in institutions of global governance such as the World Bank and the United Nations, which had five Arab founding member states in 1945: Egypt, Iraq, Lebanon, Saudi Arabia and Syria. Although the Arabic press frequently criticised the UN over Palestine and other issues, they held the organisation's ideals in high regard.[107] Moreover, despite the dominance of super powers and Western European countries, Egyptian representatives and champions of the Third World rose to important positions within international organisations. In 1983, Ibrahim Shihata became the first citizen of a developing country to be appointed senior vice-president of and general counsel for the World Bank. Boutros Boutros-Ghali served as sixth secretary general of the UN from 1992 to 1996. This was followed by Mohamed ElBaradei's tenure as director general of the International Atomic Energy Agency from 1997 to 2009.[108]

While Shihata, Boutros-Ghali and ElBaradei all had backgrounds in law, an Egyptian astrophysicist named Ibrahim Helmi Abdel-Rahman likewise rose to prominence within the United Nations. Educated in Cairo and Edinburgh, he worked at the Helwan Observatory before representing his country at the International Atomic Energy Agency in Vienna and directing the Egyptian Institute of National Planning. In 1963, Abdel-Rahman became the UN's commissioner for industrial development and, in 1966, he was made the first executive director of the United Nations Industrial Development Organization (UNIDO).[109] While serving his country and other developing nations, he was a 'cosmopolitan patriot'.[110] An obituary by a subsequent Egyptian ambassador to Vienna said that Abdel-Rahman 'had always proven that science has no homeland and no nationality, but scientists have intellectual allegiances, national attachments and human feelings.'[111]

Superpower rivalry during the Cold War also allowed an Egyptian, as a representative of a non-aligned country, to head United Nations bodies for the governance of space. Shortly after the 1957 launch of *Sputnik-1* during the International Geophysical Year, the US permanent representative to the UN requested the formation of a committee and 'Programme for International Co-operation in the Field of Outer Space'.[112] The General Assembly thus established the Committee on the Peaceful Uses of Outer Space (COPUOS) in recognition of 'the common interest of mankind in outer space' as well as the 'common aim that outer space should be used for peaceful purposes only.' Hoping to 'avoid the extension of present national rivalries into this new field', the UN considered international cooperation as a way to 'promote mutual understanding and the strengthening of friendly relations among peoples.'[113] COPUOS's membership in 1959 comprised twenty-four states, including Lebanon and Gamal Abdel Nasser's United Arab Republic.[114] In 1962, Abdel Hamid Abdel-Ghani became secretary of COPUOS and head of the United Nations Outer Space Affairs Group.[115] Over the decades the COPUOS membership expanded, thus allowing more Arab states to join. Morocco joined in 1961,[116] followed by Iraq in 1977.[117] Syria, which had seceded from the United Arab Republic in 1961, re-joined the committee in 1980. As a result, five out of the then fifty-three COPUOS member states were Arab.[118]

At the United Nations, Abdel-Ghani sought to internationalise American and Soviet space exploration for the benefit of developing countries in particular. He argued that 'the promotion of the peaceful uses of outer space should not be restricted to co-operation between East and West, but should also be extended to cover North and South.'[119] In 1966, the Egyptian diplomat published an article in the *UNESCO Courier* entitled 'Man and Space: A Great International Venture'. Almost ten years after the launch of *Sputnik-1*, he wrote, 'many nations of the world are now aware of their stake in this new activity of mankind. The nations are determined to build the necessary international structure which will ensure the peaceful exploration of outer space for the benefit of all.' COPUOS thus recommended the exchange of information, international cooperation programmes and the creation of international launch facilities for sounding rockets.[120]

The internationalism of Abdel-Ghani and COPUOS also made it into the Outer Space Treaty, signed in 1967 not only by America and the Soviet Union, but also by Egypt, Lebanon and Tunisia, among other countries. This treaty stated that the 'exploration and use of outer space, including the moon and other celestial bodies, shall be carried out for the benefit and in the interests of all countries, irrespective of their degree of economic or scientific development, and shall be the province of all mankind.' The treaty also called for 'freedom of scientific investigation' and declared that celestial bodies were 'not subject to national appropriation'. While the treaty barred states from establishing military bases and testing weapons on these bodies, it did not prohibit the 'use of military personnel for scientific research'.[121] Abdel-Ghani commented on the conclusion of the treaty, 'The jurist in the field of outer space came in advance of the scientist while in other, similar fields the scientist has always been in advance of the jurist.'[122]

Under Abdel-Ghani's leadership, COPUOS organised a United Nations Conference on the Exploration and Peaceful Uses of Outer Space in Vienna in 1968. This conference, where Abdel-Ghani served as the executive secretary, again sought to extend the benefits of space to developing countries. It resulted in the creation of the United Nations Program on Space Applications, which assisted developing nations through training, symposia and working groups on remote sensing and satellites.[123] Arab participants not only consumed, but also disseminated knowledge at the conference. During one of the evening sessions, an Egyptian scientist lectured on his country's experiences in using sounding rockets to study weather and the upper atmosphere.[124] Although Iraq was not yet a member of COPUOS at this point, it was represented by May Kaftan-Kassim at the conference.[125]

In 1974, Abdel-Ghani retired from the United Nations Outer Space Affairs Division, but continued to call for the internationalisation of space at various conferences. An American diplomat commented on his retirement that he had 'on occasion engaged in actions against US interests'.[126] However, Abdel-Ghani still received tributes, an Interkosmos medal from a Soviet representative and a copy of a NASA-commissioned book by his American counterpart.[127] In 1982, the Egyptian delivered a message by the United Nations secretary general on the 'Benefits of International Cooperation' at the conference on Remote Sensing of Arid and Semi-Arid Lands in Cairo.[128] In 1984, the former diplomat participated in a United Nations University symposium on the 'Conditions Essential for Maintaining Outer Space for Peaceful Uses'. He urged the audience to 'find new ways and new machinery to face the invasion of outer space and to keep outer space for real and sincere peaceful uses for the benefit of mankind as a whole.'[129]

A citizen of the world as well as of Egypt, Abdel-Ghani served as a senior consultant to the second United Nations Conference on the

Exploration and Peaceful Uses of Outer Space (UNISPACE II) in Vienna in 1982. Reflecting growing Arab interest, the following countries were all represented at the conference: Algeria, Egypt, Kuwait, Iraq, Lebanon, Libya, Qatar, Syria, South Yemen and Tunisia.[130] The most important theme was again the transfer of space technology to developing countries. In addition, the delegates criticised the use of space technology in military conflicts. Algeria proposed that the rapporteur's report should include a condemnation of the 'aggression committed against Lebanon and the barbarous massacres committed against the Palestinian and Lebanese civilian populations by Israel, as well as the assistance supplied to the aggressor.' However, resistance by US diplomats resulted in the removal of Israel's name from the statement.[131]

Abdel-Ghani's role perhaps contributed to the popular imagination of a cosmopolitan future of Arab space science under the control of the United Nations. In 1981, the Egyptian journalist Husayn Qadri published a utopian novel entitled *Flight into Space*. He imagined that by 1983, a 'global space centre' belonging to the UN would have brought together the space efforts of different countries 'for the benefit of mankind as a whole'. This fictional center employed 1200 scientists 'from all countries'. It was led by a board of directors consisting of twelve senior scientists, each of whom had a 'different nationality'. This center subsequently chose three Egyptians to travel to an alien world.[132]

Despite the best efforts of Abdel-Ghani, this centralisation of space activities under the control of the United Nations remained science fiction. Nevertheless, developed countries were open to international collaboration and to selling expertise and technology to Arab countries. In 1977, the Saudi Arabian National Center for Science and Technology and the US Department of Energy signed a cooperation agreement, which covered a Saudi Solar Village Project. Worth $26 million, this project used the sun and American technology to power a village near Riyadh.[133] Over ten years, it also built up a large database of solar observations.[134] When May Kaftan-Kassim requested collaboration from Germany's Max Planck Institute for Radio Astronomy in establishing the Iraqi National Astronomical Observatory, the institute's director had several reasons for responding positively. He felt that scientific initiatives in developing countries deserved support. He also saw opportunities for German industry and the possibility of recouping the growing costs by selling Iraq a replica of a thirty-metre telescope.[135]

International projects continued even during periods of intense nationalism. The Iraqi National Astronomical Observatory project initially went ahead, despite the outbreak of the war with Iran and the Israeli bombardment of the French-built Osirak nuclear reactor in 1981. During the same year, the contract negotiations were finalised for an observatory with a German consortium involving Krupp, MAN and Carl Zeiss. While the observatory was a turnkey project, Kaftan-Kassim arranged for young Iraqi astronomers and engineers to visit Germany and learn about the construction and operation of the telescopes. Only after Iranian forces had damaged the half-finished observatory, which included the radio telescope and one of two optical telescopes, did negotiations over the termination of the contract start. They were concluded with Zeiss's delivery of the other optical telescope in crates to Baghdad.[136] There they survived subsequent wars and still remained after the US invasion in 2003.[137]

While the United Nations Security Council imposed sanctions against Iraq in 1990, exchange between Western and other Arab countries in space science and technology continued. The two UNISPACE conferences contributed to a series of workshops on basic space science that were funded by the United Nations, the European Space Agency and various host countries. In the spirit of scientific internationalism, Abdus Salam, the Pakistani founder of the International Centre for Theoretical Physics in Trieste, helped establish these workshops.[138] The objective of the first, held in India in 1991, was to 'strengthen basic space science in developing countries'.[139] In 1994, the Egyptian government hosted the fourth workshop in Cairo. Five years later, the Institute of Astronomy and Space Sciences at Al al-Bayt University hosted the Eighth UN/ESA Workshop on Basic Space Science.[140]

The workshops not only brought Arab scientists together with colleagues from a variety of other nations, they also created an impetus for the development of observatories. In conjunction with the fourth workshop, the Egyptian government decided to modernise the two-metre telescope at the Helwan Observatory's Kottamia station. The refurbished telescope, which used a new Zeiss optical system, saw first light in 1997.[141] In addition, Egyptian participation with drilling technology on a Mars mission was discussed at the fourth workshop. However, this project, which was led by Mosalam Shaltout, was ultimately halted.[142] In Jordan, the eighth workshop had two follow-up projects: the operation of a small observatory at Al al-Bayt University and the conversion of a communications dish into a radio telescope at the University of Jordan.[143]

While the United Nations were important in facilitating international cooperation, transregional exchange between space scientists did not depend on large institutions of global governance. Just as space Arabism could exist without a common Arab space agency, cosmopolitanism could exist without anything close to a world state. Long before the establishment of the United Nations, North African observatories had participated in international projects. The invention of the telegraph and the loss of ships to storms encouraged the creation of a vast international network of meteorological stations. This network included an observatory in Algiers that was established in 1856. By the end of the century, the French state had invested significantly in the Algiers Observatory, which offered better viewing conditions than Paris. As one of the best-equipped French observatories, it also attracted British astronomers, including Herbert Hall Turner and Hugh Frank Newall, who observed a total solar eclipse in 1900. The observatory also became a major contributor to the *Carte du ciel*, a massive international project to map stars, which was launched in 1887. In 1926, the observatory also formed part of the global longitude operation.[144]

Transnational collaborations outside of the framework of the United Nations continued during the Cold War, even as the world seemed deeply divided. These collaborations also encompassed former enemies. The rocket programs in Algeria and Egypt – like the parallel American and Soviet ones – partly relied on German engineers who had been involved in the development of the V-2 rocket.[145] These programs were part of the 'global, transnational history' of the Nazi aerospace exodus after World War II.[146] The French press was disturbed by the fact that German was being spoken on the launching ramp in Algeria.[147] The Egyptian media were less concerned about foreign contributions to their space initiatives. In 1950, the Egyptian magazine *al-Hilāl*, meaning 'the Crescent', printed an interview with the British futurist Arthur Clarke entitled 'We will travel to the Moon'.[148] In 1959, the science fiction film *Journey to the Moon* depicted the launch of a German-built rocket with two Egyptians and one German on board.[149]

These transnational collaborations took place, even though rockets were considered a sensitive technology for national security from the 1950s onwards. Through the media, people in the Arab region had become aware of the connection between security and rocketry early on in the space age. The rocket in the movie *Journey to the Moon* has no military capability, and its three passengers are civilians. Nevertheless, the 'rocket experiment area' close to the Helwan Observatory is guarded by soldiers. Ignoring a 'photography forbidden' sign, one of the main Egyptian characters smuggles a camera into the area. Chased by the soldiers and the German scientist as a spy, he almost gets shot. However, he accidentally launches the rocket and travels to the Moon with his Egyptian colleague and the German scientist. On the Moon, they encounter 'Mister Cosmo' and his daughters and return with them to Earth. However, for their return they need to get hold of some nuclear fuel for the rocket, which is guarded by lunar soldiers.[150]

Many international collaborations also survived the intense nationalism during the decolonisation period. Participating in the International Geophysical Year, the French launched sounding rockets from Algeria to study Earth's upper atmosphere during the late 1950s. A subsequent collaboration with NASA led to the launch of FR-1, the second French satellite, in 1965.[151] After the French withdrawal, the independent Algerian state continued international collaboration in the field of satellites. It was a founding member of International Telecommunications Satellite Organization and the first developing country to use Intelsat for domestic communications during the 1970s.[152]

International exchange in Arab space science involved not just Western countries, but – transcending the Cold War divide – also the Soviet Union and its allies. Whereas the American and Soviet governments were at times mistrustful of scientific relations with one another, non-aligned Arab countries, especially Egypt, were eager to benefit from both superpowers and their allies. In 1957, the Egyptian government assisted an expedition led by, and accepted an electrophotometer from, the Academy of Sciences in the USSR. This expedition sought to observe zodiacal light at Aswan as part of the International Geophysical Year and acquaint Egyptians with 'achievements of Soviet science'.[153] In the 1970s, the Helwan Observatory first accepted lasers from the Czech Technical University in Prague as part of the Soviet Interkosmos space program. After the conclusion of the Egyptian peace treaty during the late 1970s, the observatory received American aid money for running the equipment through the Smithsonian Institution.[154] During the same decade, Soviet, Egyptian and American astronomers helped their Iraqi peers select a site for their National Astronomical Observatory.[155]

International cooperation did not remain restricted to more developed countries in the North such as America and the Soviet Union. The Iraqi rocketry program in particular also relied on South–South connections. In the 1980s, Iraq collaborated with Argentina and Egypt in developing a missile called Cóndor II in Spanish and Badr-2000 in Arabic. Although *badr* means 'full moon', this rocket was intended to strike Iran rather than explore space. As part of this collaboration, around $1 billion was funnelled through Egypt. Although this collaboration ended due to disputes and American pressure, it had provided Iraq with improved missiles called al-Ḥusayn. While this name refers to a grandson of the Prophet Muhammad and a Shia imam, it was also used ironically against the Islamic Republic. By the end of the war in 1988, almost 200 al-Ḥusayn missiles had been launched against Iranian cities.[156]

Transregional collaborations made Arab space research not just the science of other worlds such as Qatar-1b, but also fully part of global science. The asteroids 858 El Djezaïr and 859 Bouzaréah had the Arabic names of the location of the Algiers Observatory. Their discovery in 1916, however, contributed to global, rather than merely Arabic, Algerian or French astronomical knowledge. Projects transcending the Cold War divide were also more global than projects restricted to members of NATO or the Warsaw Pact.

A pertinent example is a paper on a joint Egyptian–Interkosmos–American satellite laser ranging project from 1979. This paper had twelve co-authors from the following institutions: the Helwan Institute of Astronomy and Geophysics, the Czech Technical University in Prague, the Academy of Sciences of the USSR and the Smithsonian Astrophysical Observatory.[157]

Along with an exchange of instruments, expertise and money, a global circulation of people characterised this Arab world science. The Helwan Observatory was an important station for astronomers moving between Europe and Africa. Harold Knox-Shaw, a Trinity College, Cambridge graduate, started working as an assistant astronomer at Helwan in 1907 and was promoted to superintendent of the observatory in 1913. In 1924, he left this position to become the Radcliffe Observer at Oxford. However, after having become accustomed to clearer skies in Africa, he lobbied for a new observatory in Pretoria, where he moved in 1939. Clive Langton Gregory, another Cambridge graduate, had a similar career. In 1919, he became chief assistant at the Helwan Observatory and in 1921 he started working at University College, London. After developing a strong interest in observational astronomy in Egypt, he led the establishment of the University of London Observatory.[158]

Space scientists in the Arab region moved not just for jobs, but also for education. May Kaftan-Kassim, Athem Alsabti and Hamid Al-Naimiy all completed degrees at the University of Manchester with scholarships from the Iraqi government between the 1950s and 1970s. In addition to European former colonial powers, America attracted students from the region such as Nidhal Guessoum. While most of these students returned to their home countries after graduating, many of them relocated again later in their careers. Kaftan-Kassim, for instance, moved back and forth between Iraq and America between the 1950s and 1980s.[159] Benefitting from national science funding and transnational mobility, the Iraqi scientist received support from US institutions. At the beginning of her career, she was part of the 'international community'[160] that forged radio astronomy. In the 1950s and early 1960s, after her studies at Manchester, she undertook research in radio astronomy at Harvard College Observatory together with colleagues from India and

America.[161] Between 1964 and 1966, she worked for the National Radio Astronomy Observatory in West Virginia, which operated under contract with the National Science Foundation (NSF).[162] While teaching at the State University of New York at Albany later, she won NSF grants that enabled her to research nebulae using observatories in America and the Netherlands.[163]

Transnational connections not only helped with funding and technologies, but also shaped ideologies. Various astronomers and their governments embraced internationalism throughout the 20th century. Algeria was, as part of France, a founding member of the International Astronomical Union in 1919, which incorporated the *Carte du ciel* project. Egypt joined the union in 1925, and other Arab countries followed suit in the second half of the 20th century, although their memberships ceased at times. Lebanon first joined in 1954, Iraq in 1976, and Morocco and Saudi Arabia in 1988.[164] At the IAU, Arab representatives also met their counterparts from Israel, which had joined in 1954. At the Sixteenth General Assembly in 1976, Yuval Ne'eman, who had studied under the Pakistani physicist Abdus Salam, welcomed Iraq, which was represented by Kaftan-Kassim, on behalf of Israel. On this occasion, Ne'eman emphasised scientific cooperation in the Middle East.[165]

Adding to the contributions by member states and their institutions, individual Arab members of the IAU also sought to promote their disciplines. These members were especially important for countries such as Iraq, which at times withdrew or were excluded from large parts of the international scientific community. In 2010, no Iraqi institution, but seven individual astronomers were still members of the IAU.[166] Especially active was Athem Alsabti, who had first joined the union in 1976.[167] After leaving Iraq during the late 1990s, he became part of the IAU's Program Group for the World-Wide Development of Astronomy.[168] In honour of his work, the IAU named the asteroid 10478 Alsabti after him in 2001.[169] After the 2003 war, he visited Iraq on behalf of the program group, reporting 'very strong support from the Kurdish scientists and authorities' for restoring the Iraqi National Astronomical Observatory.[170] The astronomer, then based at the University of London Observatory, hoped that the facility could become 'a center for the whole Middle East'.[171]

Arab astronomers also encountered internationalist teachers and colleagues during their stays abroad. One of the most important was Zdeněk Kopal, an 'itinerant astronomer'.[172] He first studied astronomy in Prague and Cambridge, England, during the 1930s before moving to Cambridge, Massachusetts. He first co-authored a paper with Egyptian astronomer Salah Hamid in 1950, when both were at the Harvard College Observatory.[173] The following year, Kopal moved back to England to take up a chair of astronomy at the University of Manchester. As his new base offered few days with clear skies, this gave him the incentive to collaborate with colleagues in sunnier locations, the Middle East in particular. Between the 1950s and 1980s, he attracted many students from Egypt, Iraq and Libya, as well as from Iran and Turkey.[174] During the same period, Kopal lamented that some journals had moved from an international to a national outlook. He mentioned the American Astronomical Society's *Astrophysical Journal*, which in 1963 had removed its subtitle *An International Review of Spectroscopy and Astronomical Physics*.[175]

To compensate for the nationalisation of important journals, Kopal founded new international platforms for astronomy. *Icarus*, established in 1962, had the subtitle *International Journal of the Solar System*. The journal *Astrophysics and Space Science*, founded in 1968, had a similar global mission, as expressed by Kopal in the first preface. While most other journals were serving 'their regional or national communities', Kopal argued, 'our aim will be to serve a world-wide community.' He added that his journal would be 'open on equal terms to contributions from any country – regardless of regional, national, or continental boundaries.' He established an editorial board with colleagues from twenty-six countries and solicited articles in English, French, German and Russian.[176] Through Kopal's network of students and collaborators, *Icarus* and *Astrophysics and Space Science* became important platforms for astronomical research in the Middle East. *Icarus* had several contributors who were affiliated with the Helwan Observatory during the 1960s.[177] Hamid Al-Naimiy alone published at least ten articles in *Astrophysics and Space Science*, mainly on variable stars, during the 1970s and 1980s.[178]

In addition to these journals, Kopal and his colleagues in Manchester initiated a series of International Schools for Young

Astronomers, which were supported by the International Astronomical Union and the United Nations Educational, Scientific and Cultural Organization (UNESCO). In 1967, the University of Manchester hosted the first of these schools with twelve students, including two from Egypt. Kopal himself had invited them, writing to Abdel Hamid Samaha, then director of the Helwan Observatory, 'We should not consider the school complete without the participation of some of our friends from Cairo.'[179] Manchester scientists, including Kopal, also taught at several subsequent schools that were based in developing countries, including India and Indonesia. Egypt and Morocco each hosted two International Schools for Young Astronomers between 1981 and 2004.[180]

Support from different entities, including states, regional and international organisations, thus all contributed to the growth of Arab space science. Because of the importance of relationships with colonial powers, many projects in this Arab world science had British or French participants. However, one of the earliest instances of transnational collaboration in modern Arab astronomy involved the United States, which did not formally colonise any country in the region. In the 19th century, American scientists were not yet lured to the Near East by oil, but rather by religious convictions and the moving shadows of solar eclipses.

In addition to observing eclipses, these scientists contributed to one of the oldest and longest lasting centers of modern astronomical research in the Arab region. This center was the Lee Observatory of the American University of Beirut (AUB), which itself formed one of the first modern institutions of higher education in the Middle East. The next chapter will focus on the global networks that sustained this observatory for around a hundred years, from the 1870s to the 1970s. This period also saw the emergence of Arab nationalism, which likewise made AUB one of its centres. The next chapter will explore how and to what extent this nationalism as well as the university's connections with America and Europe shaped the beginnings of modern astrophysics in the Arab region.

# CHAPTER 3

# East Coast and Near East Colleges

In the early 1900s, David Todd, an astronomer at Amherst College in Massachusetts, travelled twice to Tripoli in North Africa. His goal was to observe two total solar eclipses occurring over the same city within a few years, a rare coincidence. During both trips, Todd operated a self-designed forerunner of the movie camera to document the eclipses.[1] In May 1900, an automated mechanism using a commutator made more than a hundred exposures during the fifty-five seconds of totality.[2] In August 1905, the astronomer secured sixty-three 'fine pictures' of the corona in three minutes, as he later reported in the journal *Science*. The pictures revealed 'coronal streamers to exceptional length'[3] and proved that the type of corona observed was 'inseparably related to a thickly spotted sun'.[4]

Todd's eclipse expeditions to Tripoli were not only a scientific venture, but also an experience in cosmopolitanism. Accompanying him was his wife Mabel, a writer, editor of Emily Dickinson's poems and lover of the poet's brother Austin. Mabel wrote that a daughter of the British consul-general 'illustrated the true cosmopolitan, at home in the world at large'. She added, 'many an afternoon have I seen her dispensing tea to half a dozen nationalities, addressing each in his own language, turning instantly and gracefully from one to another with English, French, Turkish, Italian, Arabic, modern Greek, Maltese – on the tip of her clever and fluent tongue.' This cosmopolitanism was not restricted to the city's European elites. Mabel stated that 'daily life presented a moving picture of nationalities well-nigh unmatched by any other region.' She mentioned that 'even the small boys of Tripoli's narrow streets can use interchangeably all the languages current there.' During the eclipse of 1900, Mabel observed that 'every roof all over the city was swarming with humanity, Maltese, Jews, Arabs, Turkish soldiers on their upper ramparts, Franciscan monks on their high-air promenade – even the minarets were crowded.'[5]

The Todds not only admired the peaceful interaction of people from various backgrounds; they also depended on it for their scientific work. In the absence of an American consulate, David and Mabel set up their observing station on the roof-terrace of the British consulate-general, which was soon called the 'royal observatory'. The Ottoman governor forbade fires and ordered the sprinkling of the streets to prevent smoke and dust from obscuring the eclipse.[6] Arabs and Maltese constructed the commutators, incorporating Islamic prayer beads into the mechanism.[7] Telegrams from America described the eclipse before its arrival in Tripoli, while an Italian drawing-master and his pupils were getting ready to sketch the corona.[8] Demonstrating cosmopolitan skills themselves, the Todds turned many of their collaborators into friends. When Mabel published a book about Tripoli in 1912, she dedicated it 'to them all, loved and remembered, English, Turkish, Italian, French, Arab'.[9]

Although the eclipses were fleeting, the Todds' trips to Tripoli were not merely a 'social event of the season'.[10] Instead, they were part of a longer circulation of astronomers across the Atlantic and the Mediterranean. What attracted American astronomers like Todd to North Africa was that eclipses were global phenomena as well as phenomena that could best be observed in certain places. As the Earth turned, an eclipse's shadow moved around the globe, irrespective of national boundaries. In order to observe the eclipses, astronomers had to move to whichever world region the shadow band crossed. Because North Africa had many days with clear skies, it was especially promising for observations. Between their two trips to Tripoli, the Todds went on another eclipse expedition to the Dutch East Indies in 1901, although in this tropical location clouds obscured the sun during the six minutes of totality,[11] bringing 'disaster' for the solar photographers.[12] Later in his life, Todd confessed to his daughter that what he regretted most was not Mabel's affair with Emily Dickinson's brother Austin, but 'three cloudy eclipses of the Sun'.[13]

Facilitating the global circumnavigation of some astronomers such as Todd were networks between colleges on America's East Coast, including Amherst, and in the eastern Mediterranean, in particular the Syrian Protestant College (SPC) in Beirut. In 1899, Raymond Dugan, one of David Todd's former students at Amherst, became an instructor in mathematics and astronomy at the SPC.[14] The two remained in contact, with Todd requesting that Dugan and the SPC's twelve-inch telescope join him on his expedition to the Dutch East Indies.[15] The SPC's leadership rejected this particular proposal due to its cost, but Dugan remained internationally mobile.

Travelling between the US, Europe and the Near East, astronomers affiliated with the SPC such as Dugan contributed to world science. After three years at the SPC, Dugan returned to Amherst for a master's degree, but immediately left to earn a doctorate at Heidelberg under Max Wolf. There, he discovered eighteen minor planets,[16] naming three of them after the Todds and his alma mater: 510 Mabella, 511 Davida and 516 Amherstia.[17] Another asteroid, 518 Halawe, expressed his connection to Beirut.[18] Reportedly, Dugan 'was over-fond of these Arab sweetmeats, and occasionally required medical assistance because of them.'[19] In 1905, Dugan contributed to observations of the same eclipse that Todd was studying, but from Spain instead of Tripoli.[20] Following the completion of his doctorate, he returned to America, where he spent the rest of his career at Princeton University and became a leader of his profession. He served as secretary and vice-president of the American Astronomical Society. From 1935 until his death in 1940, he also headed the International Astronomical Union's Commission on Variable Stars.[21]

The connections between East Coast and Near East colleges remained strong, surviving the various wars that affected Lebanon during the 20th and 21st centuries. Renamed the American University of Beirut in 1920, the Syrian Protestant College formed a hub for astronomical research in the Arab region, as it did in social research.[22] Made up of faculty and students from diverse backgrounds, it remained cosmopolitan despite the growth of Arab nationalism and later the outbreak of the Lebanese Civil War.[23]

Indeed, Lebanon's later wars only increased migration and the circulation of researchers between North America, Europe and the Middle East. Thanks to this circulation, many researchers continued to be part of world science, while often cultivating a special interest in the history of Arabic astronomy.

## COLLEGIATE CONNECTIONS

The growth of AUB as a centre of modern astronomical research in the Near East depended significantly on connections with institutions in America's northeast, especially Amherst. Established in 1821, Amherst College was itself an institution with a transnational mission. Its founders had the ambition to educate men for 'civilizing and evangelizing the world'.[24] Many of its graduates served as missionaries, including Daniel and Howard Bliss, the first two presidents of the SPC, as well as George Washburn, the second president of Robert College in Istanbul. In preparing its students for religious and educational service, Amherst College sought to combine science and religion. When Daniel Bliss was studying at Amherst, its president, Edward Hitchcock, held a chair of natural theology and geology. Hitchcock was also the author of a book entitled *The Religion of Geology and Its Connected Sciences*.[25]

Amherst served as a model for the Syrian Protestant College in combining modern astronomical research with proselytism. Under Hitchcock, Amherst's first observatory was constructed in 1847 with donations from Massachusetts philanthropists.[26] 'We cannot dissever education and Christianity,' one of Amherst's Yale-educated trustees stated during the dedication of the observatory, 'Remember the commission: *Go ye into all the world.* Improve the condition of man.'[27] This observatory subsequently attracted David Todd away from Columbia College in New York to study at Amherst, where he graduated in 1875.[28] In planning the establishment of the SPC, Daniel Bliss wrote to one of his former professors, 'If the Institution is favored of God & of man it shall be the Amherst – or the Yale of the Orient.'[29]

Although the Syrian Protestant College was located in the Ottoman Empire, it had an extraterritorial character and – like Amherst – strong connections to Protestant philanthropists. In 1862, the American Board of Commissioners for Foreign Missions charged Daniel Bliss with founding 'a college of higher learning that would include medical training; this college should have an American educational character.'[30] As the funds for running the institution were to be raised in America and Britain, it was deemed necessary to have an American or British board of trustees 'to handle the funds and lend prestige to the undertaking so that large contributors might have confidence.' Similarly, the first donors felt that the college should have a British or American charter as a 'proper legal basis'. Bliss noted that 'it probably never occurred to any one even to ask the question whether such a Board of Trustees should be legalized by the Turkish Government.'[31] In 1863, the board was incorporated in New York, with William Dodge, a wealthy businessman, serving as its treasurer. The following year, Dodge's son Stuart was appointed the first professor of modern languages, and the college officially opened in 1866. In 1871, William Dodge himself came to Beirut in order to lay the cornerstone of the new College Hall.[32] This building was planned by a New York-based architect who also designed college buildings in Princeton.[33]

Although modelled after institutions on America's East Coast, the Syrian Protestant soon became more diverse and cosmopolitan. The college's motto was taken from the Bible: 'that they may have life, and have it abundantly' (John 10:10). By teaching medicine and other subjects, the SPC aimed at providing relief and prosperity for all people in the Near East – and to attract them to Protestantism. During William Dodge's visit, President Daniel Bliss stated: 'This College is for all conditions and classes of men, without regard to color, nationality, race or religion.' Bliss added, 'A man, white, black, or yellow; Christian, Jew, Muhammedan or heathen, may enter and enjoy all the advantages of this institution.'[34] As in instances of colonial cosmopolitanism during the period, this openness did not mean an absence of racial hierarchies. Bliss later wrote in his *Reminiscences*, 'We open its doors to the members of the most advanced and the most backward of races.'[35] Despite racism, however, students from different backgrounds were indeed admitted. After initially mainly recruiting orthodox Christians, the college's student body soon diversified to a greater extent than its main competitor in Beirut, the Jesuit Saint Joseph University, and many colleges in America.[36] By 1908, the student body of 827 included 147 Protestants, 127 Muslims, sixty-two Jews and twenty Druze.[37]

Within this cosmopolitan setting, modern astronomical work took place almost from the beginning. The SPC soon gave the discipline a similar status as Amherst College did, where the teaching of astronomy was first recorded in 1825, within a few years of its establishment.[38] The first instructor of the subject at the SPC was Cornelius Van Dyck, a medical doctor from New York. Caught up in a wave of missionary enthusiasm, he offered his services to the American Board of Commissioners for Foreign Missions, which sent him to Syria. After his arrival in 1840, he mastered Arabic with the help of Butrus al-Bustani and Nasif al-Yaziji.[39] Together with both of them, he formed part of an international network of scholars who collaborated in translating the Bible into Arabic.[40] In 1852, he also published an Arabic textbook on geography.[41] After Van Dyck's death in 1895, a sheikh from al-Azhar told the missionary's daughter, 'Your father taught me, by his published writings, that it is possible to write good Arabic, correct in grammar and in idiom, in a style so simple and so clear as to be easily understood by any intelligent reader, whether learned or unlearned.'[42]

Van Dyck also developed the skills of a cosmopolitan, developing the ability to interact with people from the Near East as well as America. Over the years, he integrated into local society to such an extent that it once almost cost him his life. He adopted a Syrian costume, including a red fez, Turkish trousers and slippers turned up at the toes. During a period of conflict in Mount Lebanon, armed Druze once mistook him for a local Christian. Van Dyck tried in vain to convince them that he was a foreigner, an American doctor who cared for, and was ultimately interested in converting, members of all religions. 'We know that no foreigner ever spoke Arabic as you do,' the Druze leader responded. Van Dyck was saved by the arrival of other Druze who recognised him as the doctor who had previously dressed the wounds of one of theirs.[43]

While Van Dyck adopted local attire, Van Dyck was critical of Arab science. Like Bliss and other American missionaries in Syria, he may have been a cosmopolitan, but he was also condescending. In 1849, he published an article 'On the Present Condition of the Medical Profession in Syria' in the *Journal of the American Oriental Society*. 'Although great praise may justly be given to the Arab nation as the preservers of science,' Van Dyck wrote, 'they deserve none as discoverers.' Unsurprisingly for a missionary, he considered Islam as an obstacle to science. 'Islamism, in itself considered, must be regarded as a desolating superstition,' he wrote. 'The same principle which led the Khalifeh 'Omar to order the burning of the Alexandrian library, has since then worked the ruin of many a fair structure, and given the death blow to many a worthy enterprise.' Under the Ottomans, 'improvement' only began, 'when their religion began to lose its hold upon their minds'.[44]

Van Dyck sought to fight ignorance by teaching Western science associated with Protestantism. After having incorporated a view of 'evangelical modernity', he and other missionaries thus spread the 'gospel of science'.[45] Based on his medical skills and knowledge of Arabic, in 1867 he was appointed professor of internal medicine and general pathology at the Syrian Protestant College. He also taught chemistry until the appointment of Edwin Lewis, another graduate of Amherst College, as professor of geology and chemistry in 1870. Thereafter, Van Dyck offered courses in astronomy and meteorology, which were his hobbies. In order to support his teaching, the SPC constructed an observatory in 1874 and extended it in 1880. It was known as the Lee Observatory, after a merchant from Manchester who had donated £150 for the building. Van Dyck purchased most of the equipment from the income from his medical practice and loaned it to the college.[46]

Through his teaching and writing, Van Dyck propagated heliocentric views in Arabic lands. In 1874, he published an Arabic book on *Principles of Astronomy* in Beirut with the permission of the Ottoman authorities. In the introduction, he asserted that 'the true system of astronomy is Copernicus's system.' Its first principle is Earth's rotation around its axis. The second is that the sun is the centre, with Earth revolving around it, 'contrary to the ancients' claim of Earth's fixity in the centre'.[47] The observatory was intended to further fight traditional beliefs. In 1877, Daniel Bliss wrote in a report to the board that the observatory would 'prove useful in the direct education of students and in attracting the attention of natives to the superiority of Western knowledge, thus helping to dispel ancient deep-rooted superstitions.'[48]

Van Dyck also spread modern techniques of observation and consciousness of a global scientific community by working with former students. In 1873, the SPC hired Yakub Sarruf, a member of the first graduating class, as a 'native tutor' in Arabic, mathematics and physics. A year later, the doctor recruited the recently graduated Faris Nimr to assist him in the newly established observatory. Together, the scientists formed part of a global community that observed the two transits of Venus during the 19th century. During the 1874 transit, he, Van Dyck and the doctor's daughter 'did what others did at many observatories scattered over the face of the Earth', Nimr later remembered.[49] Back in America, David Todd used various American photographs of this transit to calculate the distance between Earth and the sun as being equal to around 148 million kilometres.[50]

Through the participation of the global community, successful observations in one location could compensate for failures in another. In 1882, the next transit of Venus again brought excitement for Yakub Sarruf and Faris Nimr, but also disappointment. Conscious of the event's global reach, the two scientists wrote that 'observation has been divided between the countries of the Earth' in order for astronomers to 'verify the distance of the sun'. 'For the love of science and to establish knowledge of the basis of truth,' astronomers were spending 'huge sums of money' and accepting 'hardship and risks'. However, Sarruf and Nimr also lamented that 'years on the path of knowledge' had sometimes earned scientists 'mockery' and 'animosity'.[51] Indeed, even David Todd later became better known as a 'famous cuckold than astronomer'.[52] Despite their preparations, Sarruf and Nimr were unable to observe the 1882 transit, because of clouds over Beirut.[53] Todd had more luck at California's Lick Observatory. 'Splendid day. Splendid success', he telegraphed Mabel. He took more than 140 photographs, which the Princeton astronomer Charles Young called 'probably' the best obtained anywhere.[54] In 2004, astronomers from the Lick Observatory described the plates as 'the most complete photographic record of the transit' and, with assistance from the magazine *Sky and Telescope*, turned them into a movie.[55]

While the Syrians Protestant College familiarised its students with global astronomical research through hands-on practice, it reached many more Arabs through its publications. With Cornelius Van Dyck as their mentor, Yakub Sarruf and Faris Nimr started publishing an Arabic scientific periodical in 1876.[56] Van Dyck suggested its name, *Al-Muktataf*, meaning 'Selections' or 'Digest', and recommended that it publish Arabic translations of extracts from Western scientific journals. Van Dyck also arranged to have the American Mission Press in Beirut publish the journal and helped secure a permit from the Ottoman director of publications.[57] The first volume of the journal contained nine articles on astronomical topics, including the moon and the solar system.[58] These articles predicted a partial and a total lunar eclipse in Syria in 1876 and 1877, and caused considerable excitement among its readers.[59] In addition, the first issue included an article on 'Astronomers among the Arabs' from the Abbasid caliph al-Maʾmun to the transfer of knowledge to Western Europe.[60]

While the Syrian Protestant College remained a cosmopolitan institution, a beleaguered Ottoman government also contributed to the emergence of nationalism in *Al-Muktataf*. Conscious of tight censorship under Sultan Abdul Hamid II, Sarruf and Nimr justified the establishment of the periodical in national terms. These terms were ambivalent enough to refer to both Syria and the wider Ottoman Empire. In their introduction to the first issue, the editors justified their venture as a 'service to the nation [*al-waṭan*]' and as a 'response to the requests by many lovers of progress'. Sarruf and Nimr added that this nation was in need of 'science and industry'. They proudly mentioned the permit from an Ottoman official who 'was known for his zeal in benefitting the nation'.[61]

A sense of participating in a global community co-existed with this vague nationalism in *Al-Muktataf*. In the second issue, Sarruf and Nimr expressed pride in the Lee Observatory's network. 'The relations between the Syrian Observatory and European and American observatories, like the Sultan's Observatory and the observatories of Vienna, London and Washington, delight the sons of the homeland.' Twice a day, the Lee Observatory transmitted meteorological data to

Istanbul, and once a day to the remaining observatories. Thus, 'everybody in America knows our weather conditions every day.' Combining Ottoman subjecthood and Syrian identity with a global perspective, Sarruf and Nimr added that this information saves 'the world great losses from shipwrecking and disruptions of livelihood'.[62]

As a pioneering Arabic scientific periodical, *Al-Muktataf* also became a platform for debating geocentric and heliocentric world-views – and thus for struggles over authority. In 1876, Yakub Sarruf published an article in which he criticised a fellow Christian for defending the notion of a fixed Earth. This criticism incited a Greek Orthodox archimandrite to publish a response that insisted on Earth's fixity based on verses in the Old Testament. Subsequently, ʿAbd Allah Fikri, the Egyptian deputy minister of education, intervened in the debate. He affirmed Earth's mobility and claimed that it could be found among the writings of earlier Muslim scholars. This led the archimandrite to accuse the owners of *Al-Muktataf* of having solicited Fikri's article. The archimandrite then quoted verses from the Qur'an that supported the fixity of Earth in his opinion. At this point, the owners of *Al-Muktataf* reminded the archimandrite that the Bible itself was to be read figuratively. They also told the Orthodox leader that his proclaimed opposition between heliocentric astronomy and scripture was not the official position of his own church. This high tone quieted the archimandrite.[63]

In the early 1880s, however, another debate on science and religion caused a setback for research at the Syrian Protestant College, including astronomical work. In 1882, the year of Charles Darwin's death, Edwin Lewis gave a commencement speech that praised the naturalist and thus 'smacked of Darwinism'. This speech was subsequently printed in *Al-Muktataf*. This came after Lewis had already caused a small 'scandal' by serving alcohol at a party in his house earlier that year. Lewis's behaviour divided the college. Van Dyck and others were on Lewis's side, whereas Daniel Bliss and George Post, the professor of botany and surgery, were against him. Importantly, Stuart Dodge, who had become secretary of the Board of Trustees, considered Lewis a 'dangerous man and unsuited to occupy a post in the College'.[64]

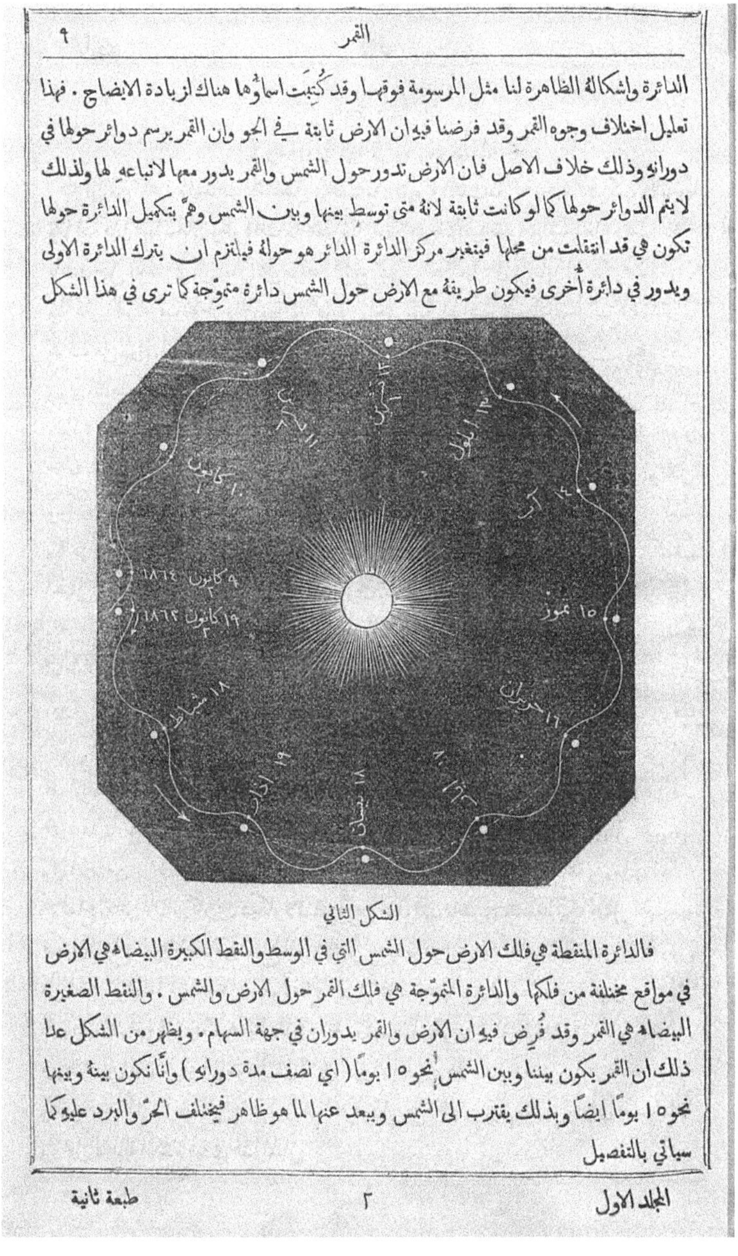

٩ القمر

الدائرة واشكالهُ الظاهرة لنا مثل المرسومة فوقها وقد كُتبت اسماؤُها هناك لزيادة الايضاح . فهذا تعليل اختلاف وجوه القمر وقد فرضنا فيهِ ان الارض ثابتة في الجو وان القمر يرسم دوائر حولها في دورانهِ وذلك خلاف الاصل فان الارض تدور حول الشمس والقمر يدور معها لانباعهِ لها ولذلك لا يتم الدوائر حولها كما لو كانت ثابتة لانهُ متى توسط بينها وبين الشمس وهمَّ بتكميل الدائرة حولها تكون هي قد انتقلت من محلها فيتغير مركز الدائرة الدائر هو حولهُ فيلتزم ان يترك الدائرة الاولى ويدور في دائرة أُخرى فيكون طريقهُ مع الارض حول الشمس دائرة متموجة كما ترى في هذا الشكل



الشكل الثاني

فالدائرة المنقطة هي فلك الارض حول الشمس التي في الوسط والنقط الكبيرة البيضاء هي الارض في مواقع مختلفة من فلكها والدائرة المتموجة هي فلك القمر حول الارض والشمس . والنقط الصغيرة البيضاء هي القمر وقد فُرض فيهِ ان الارض والقمر يدوران في جهة السهام . ويظهر من الشكل على ذلك ان القمر يكون بيننا وبين الشمس نحو ١٥ يوماً ( اي نصف مدة دورانهِ ) وانَّا نكون بينهُ وبينها نحو ١٥ يوماً ايضاً وبذلك يقترب الى الشمس ويبعد عنها لما هو ظاهر فيختلف الحرّ والبرد عليهِ كما سيأتي بالتفصيل

المجلد الاول ٢ طبعة ثانية

**Figure 3.1** Illustration of Earth moving around the sun in the first volume of *Al-Muktataf*

The Lewis affair led to the departure of many of the college's faculty, including Van Dyck, and to a serious disruption to teaching and research during the academic year 1882–83. Lewis was forced to leave, causing Van Dyck and other professors to resign in protest. Many students went on strike and presented petitions to Daniel Bliss. George Post tried to prevent students who would not repent from sitting the imperial medical examination in Istanbul. His attempt failed, but amidst the chaos, the college's president and Board of Trustees remained firm. 'I love him and shall always cherish his memory,' Stuart Dodge wrote to Bliss about Van Dyck. However, the 'terror of more resignations and wholesale desertions,' he added, 'do not and cannot disturb us.' Dodge insisted that the college continued. 'We can close it entirely for a year or two if necessary, but it is to go on.'[65]

The Lewis affair affected not just astronomical research, but scientific publishing too. In order to cover for the resignations, Yakub Sarruf and Faris Nimr, who had printed Lewis's speech in *Al-Muktataf*, were initially retained and appointed adjunct professors in chemistry and mathematics. However, suspicions of anti-Ottoman activities and their relationship with Lewis and Van Dyck jeopardised Sarruf and Nimr's positions.[66] What is more, the editors dedicated further articles to evolution and went on to write about Darwin as 'the most learned of this age and the most famous of its men'.[67] In 1884, the Board of Trustees thus terminated their employment while still 'commending them for their ability and assiduity during their long service'.[68] In order to avoid Ottoman censorship, they continued to publish *Al-Muktataf* from Cairo. Dodge commented on their departure, 'I do not feel troubled about Nimr and Sarruf as far as the College is concerned, but those young men will turn miserably unless grace prevents. "Going down to Egypt" has never proved wholesome.'[69]

In contrast to Dodge's predictions, Sarruf and Nimr's publishing enterprise flourished in Egypt and expanded beyond the scientific fields. As Arab patriots and Anglophile cosmopolitans, they established strong connections with the elites of a country that had been occupied by the British in 1882. In 1888, Nimr married the daughter of the British consul in Alexandria, Helen Eynaud, whose ancestors included French and Austrians. They promoted Egyptian and Syrian independence from the Ottoman Empire and hoped the Arabic lands would develop under British tutelage. In addition to *Al-Muktataf*, Sarruf and Nimr founded *Al-Mokattam*, a daily newspaper, in 1889. It was named after a hill near Cairo and site of a Fatimid observatory constructed for the astronomer Ibn Yunus. *Al-Mokattam* was simultaneously pro-British and supportive of the Egyptian monarchy.[70]

In the liberal publishing environment of Cairo, Nimr and Sarruf paid more attention to social, cultural and political issues, but still sought to disseminate modern astronomical knowledge as well. *Al-Muktataf* continued to publish articles on topics such as the moon, Sirius, Halley's Comet and variable stars.[71] In 1891, Nimr visited Milan's Brera Observatory and asked its director, Giovanni Schiaparelli, about his claim to have discovered *canali* on Mars.[72] While some scientists translated *canali* as 'channels', others, such as Harvard alumnus Percival Lowell, caused sensation by speaking of 'canals'. In the decades after the opening of the Suez Canal in 1869, these canals denoted great feats of engineering and thus intelligent life on another planet.[73] In 1920, *Al-Muktataf* also became one of the earliest Arabic periodicals to discuss Albert Einstein's theory of relativity.[74] In addition to covering debates involving Western science, *Al-Muktataf*'s editors were proud of the Islamic scientific tradition, but did not see it as living culture or practice.[75] They thus contributed to narratives of decline in Islamic science, but also to the memory of an Arab Golden Age as a symbolic resource for subsequent generations of scientists.

Although Nimr and Sarruf never returned to Beirut as instructors, they maintained connections with East Coast and Near East colleges. The transatlantic networks that had nurtured modern Arab astronomy thus endured. During the late 1880s, Andrew Dickson White, who had just stepped down as the first president of Cornell University, met Nimr on a visit to Egypt and described him as an 'especially interesting man'.[76] In 1890, the Syrian Protestant College recognised the achievements of their two alumni by awarding them its first honorary doctorates through the State University of New York.[77] In 1926, the college, by then renamed the American

University of Beirut (AUB), hosted a party celebrating *Al-Muktataf*'s 'golden jubilee'. One of the speakers was Fuad Sarruf, a nephew of Yakub who had graduated from the institution in 1918 and subsequently taught English there. He edited the periodical from the 1920s onwards, while also teaching history at the American University in Cairo.[78]

Under the AUB graduate Fuad Sarruf, *Al-Muktataf* retained its dual cosmopolitan and nationalist character. In his jubilee speech at the university, Sarruf emphasised the periodical's importance in connecting the Arabic lands with Europe and North America. 'The journal has been a faithful messenger between the civilisations of the East and West,' he stated, bringing together writers of 'different races and beliefs'. Fuad further explained that *Al-Muktataf* was 'Eastern in its origins and purpose, Western in its style and approach.' Although a cosmopolitan, Fuad also revealed an inclination towards Arab nationalism. His periodical 'includes the sons of the East in a close moral unity' during a time of 'division', the editor stated a few years after the breakup of the Ottoman Empire.[79] Fuad retained his cosmopolitanism and Arab nationalism for the rest of this life. In 1950, he published an article entitled 'We the people of the world', in reference to the first words of the United Nations Charter. He acknowledged that the UN General Assembly 'had failed us in Palestine'. However, he still called upon Arab countries to 'unify their word', to support the UN, and to work through its institutions.[80]

## ECLIPSE EXPEDITIONS

Although the Syrian Protestant College lost Van Dyck, Nimr and Sarruf as instructors and editors during the 1880s, the strong connections with American East Coast institutions allowed for a continuation of modern astronomical teaching and research. The recruitment of new faculty members from America was helped by a change in the SPC's language of instruction from Arabic to English in 1883. One of the new instructors was Robert West, a Princeton graduate whose teacher Charles Young also observed eclipses and worked on spectroscopy of the sun. Appointed in 1883, West was supposed to replace an English instructor who had become ill.

However, when the latter recovered, West became an instructor in mathematics and astronomy instead of Nimr. In order for him to undertake observations, the college bought Van Dyck's astronomical instruments for £500.[81] In 1887, George Post was sent to America to raise $100,000 for the college's endowment fund. $30,000 would be used for a professorship in mathematics and astronomy, with the remainder being spent on the library, scientific collections and scholarships. Post presented this plan to the college's trustees at a meeting in December, and one of them immediately pledged $10,000. In the summer of 1888, Post returned to Beirut with around $57,000, enough for the professorship, to which West was promoted.[82]

Throughout his time in Beirut, Robert West followed some of the most important Western scientific journals and added to the work of his colleagues based in other parts of the world. In 1885, he published a letter to the editor in *Nature*, reporting 'a magnificent meteoric shower'. This was part of a series of letters from observers of the same phenomenon, which also included one from Charles Young. Based further east and south than many European and American observatories, he was able to offer a different view of the meteoric shower. His letter thus complemented letters from his peers in America, Britain, Switzerland and Istanbul.[83]

In Beirut, West also learned Arabic and became interested in the history of astronomy in the region, although when reading about the subject he relied mainly on previous European scholarship. In 1895, he published an article entitled 'Pronunciation of Star Names Derived from the Arabic' in the American magazine *Popular Astronomy*. West stated that the 'Arabian astronomers were the connecting link between Greek astronomy and that of the Middle Ages in Europe. The *Megala Suntaxis* of Ptolemy was the basis of their system, and the Almagest, as they called their translation, was one of the principal bases on which the science was founded in Europe.' West then gave a list of Arabic star names and their etymology and meaning, including Achernar, 'the end of the river', and Rigel, 'foot'.[84]

In order to keep pace with astronomical advances elsewhere, West sought the installation of a twelve-inch telescope and other instruments. For this new equipment, he decided in 1891 to rebuild the Lee Observatory using international expertise. His teacher Charles Young approved the design of the building, and the Grubb Telescope Company in Ireland, which also equipped the Royal Observatory in Greenwich, built the dome. The telescope's manufacturer was the Warner and Swasey Company in Cleveland, which built instruments for the Lick Observatory, the US Naval Observatory and the Yerkes Observatory. Stuart Dodge and other American philanthropists covered the new observatory's costs of around $9,500,[85] with construction starting in 1892 and ending in 1894.[86] The observatory soon enabled West to make detailed observations of variable stars. Between 1896 and 1898, he published seven articles in the *Astronomical Journal*, again complementing knowledge produced by his peers elsewhere. As the Lee Observatory's latitude was lower than those of almost all the observatories in Europe, he focused on southern stars. In this research, he benefited from the increasing observations of stars in the southern hemisphere during the 19th century. In particular, he confirmed and complemented the data in the Córdoba *Durchmusterung*, a catalogue of stars surveyed from a new observatory in Argentina.[87]

Besides studying variable stars, West used the college's new telescope to study birds, thus continuing one of Charles Young's interests. In 1880, while West was studying at Princeton, Young was involved in observing birds as they flew against the background of the moon. His telescope's angle and focus provided the data that allowed him to calculate the birds' elevation.[88] West perhaps remembered Young's research when he read an article entitled 'A Remarkable Flight of Birds' in *Nature* in 1895. This article described an observation of birds passing 'like shadows' in front of the sun. The observer, who was based in Britain, hoped 'that other astronomers, who may have seen a similar thing, may be led to mention the fact.'[89] Three months later, West published his own account in the journal. While observing the occultation of stars by the Moon in the Pleiades one evening, West estimated to have seen around 250 birds and calculated their altitudes.[90]

From the late 1890s onwards, West assumed more administrative responsibilities, but teaching and research in astronomy continued at the SPC. At the turn of the century, astronomy was required for every senior student, for the reason that 'the heavens declare the glory of God' (Psalm 19:1).[91] Between 1899 and 1903, West became principal of the preparatory department, which was said to have 'occupied all his time'.[92] In addition, West served as the college's treasurer and, in 1905, became dean of the Faculty of Arts and Sciences.[93] In the meantime, the Lee Observatory was first taken over by David Todd's student Raymond Dugan and later by Alfred Joy, a graduate from Oberlin College in Ohio. After arriving in 1904,[94] Joy also became involved in an astronomical venture organised by the Lick Observatory which consisted of three simultaneous expeditions to Labrador, Spain and Egypt. The goal was to observe the solar eclipse of 1905, which Todd recorded in Tripoli and from three other locations. The event was calculated to occur in August. This month promised clear skies in the northern hemisphere and freedom from teaching for many college-based astronomers. The eclipse thus offered a rare opportunity to determine the existence of motion in the solar corona and planets between the sun and Mercury. William Crocker, a prominent Californian banker and philanthropist, agreed to fund the expeditions.[95]

Although the Lick Observatory expeditions were organised by Americans, they would not have succeeded without the participation of people from other countries. In April 1905, *Al-Muktataf* urged the 'director of the Khedivial Observatory to go close to Aswan to monitor the eclipse from there.'[96] In addition, the British-led Survey Department, which was responsible for the Khedivial Observatory, provided vital logistical support for foreign astronomers. Even before the American team reached the African coast, the department arranged for quarters in Aswan's Savoy Hotel and delivered lumber and cement for the establishment of a camp on the hotel grounds. Henry Lyons, the department's head, met the Americans as well as two parallel British and Russian expeditions at their ports of entry. The Survey Department also paid for rail transportation and provided fruit and drinks for the stretches of desert between Cairo and Aswan. In addition, the department supplied distilled water from Cairo for the development of photographic plates.[97]

The colonial support for this multinational research on eclipses may be partly explained by nationalism. It provided British officials with an opportunity to display their science and administration to Americans and the world.[98] However, despite potential nationalist motivations, the result was a transnational venture. With the major costs of the Lick Observatory expedition to Egypt being covered by other sources, participation was attractive for scientists from the Syrian Protestant College. Robert West and Alfred Joy joined their peers in Cairo and travelled with them to Aswan, where they directed some of the observations. Participants in the eclipse expeditions also enjoyed a cosmopolitan spirit. The Lick Observatory team in Spain was joined by astronomers from Germany and Sweden,[99] while their colleagues in Egypt worked alongside their British and Russian peers. As the shadow of the eclipse approached Aswan, ministers and diplomats arrived from Cairo by boat and train. 'A startling atmosphere of the Great World breezes through the halls' of the Savoy Hotel, wrote Ethel Hussey, the wife of the Lick Observatory expedition leader. The astronomers also kept their colleagues on other continents in mind. 'We think of Labrador and our colleagues on its misty shores,' Hussey continued, 'We watch the hour pass and know that the shadow is sweeping across the chequered skies of "Sunny Spain."',[100] *Al-Muktataf* also reported on the expeditions from and to different nations.[101]

This cosmopolitan atmosphere and planetary consciousness did not mean an absence of imperial hierarchies. As Western astronomers depended on colonial logistical networks, they did not leave the world of imperialism.[102] Native participants in astronomical research included not just the editors of *Al-Muktafaf* and other members of the professional and middle classes or *effendiyah*, manual labourers were, as part of the 'colonial scientific working class',[103] vital to the success of the Lick Observatory expeditions. William Hussey wrote that 'labor was cheap and abundant at Aswan. While our colleagues in the Labrador were finding it impossible, on account of the salmon run, to get a man at any price, we could have had any number at twenty-five cents a day.' Hussey added that 'the custom of the country is to work through native overseers,' which the Survey Department provided. These intermediaries selected watchmen, labourers and contractors to build the camp.[104]

The Survey Department and Robert West with his cosmopolitan skills facilitated communication between the American and British astronomers and the Egyptians. William Hussey acknowledged that 'in addition to his technical skill, his mastery of Arabic and his insight into Oriental character were no small factors in establishing the cordial relations with our native helpers, which continued to the end of our stay.'[105] As a result, other astronomers could operate in Aswan without knowing much Arabic. Herbert Turner, the leader of the British team, thanked the Survey Department for providing men who talked 'Arabic for us (and to do a great deal of other work incidentally), though most of us have learnt the Arabic for water and ice, and practice the use of these words constantly.' Turner added, 'We have other drinks, but these can usually be obtained by mentioning their English name.'[106]

Thanks to the Egyptian weather and the multinational support from Robert West, the colonial authorities and local workers, the Lick Observatory expedition to Aswan was successful. Whereas clouds blocked the views for the Lick Observatory teams in Labrador and Spain, the sky was clear over Upper Egypt. The astronomers obtained nineteen photographs of the inner corona and the fainter outlying streamers. Packed in boxes with English, French and Arabic labels, the photographs were shipped to the Lick Observatory for further study. William Hussey wrote that as West sailed back to Beirut, he left his peers 'conscious of the loss of his resourcefulness, his congenial companionship, and his invaluable aid in all ways during our association at Aswan.'[107]

West died in 1906, soon after his return to Beirut, but he also had cosmopolitan successors. Alfred Joy became acting director of the Lee Observatory and, in 1911, was appointed director and professor of astronomy.[108] From Beirut, the astronomer and his research circulated between the Near East, Europe and North America. In 1908, he published an observation of the transit of Mercury in the American magazine *Popular Astronomy*.[109] Between 1909 and 1911, Joy worked as Turner's assistant at Oxford and as a fellow at Princeton. In 1914, Joy was at the Potsdam Astrophysical Observatory in Germany[110] and published on variable stars in *Astronomische Nachrichten*.[111] From there, he moved to the Yerkes Observatory in

Wisconsin to spend a year as an instructor and collaborator on several research programs. He planned to return to Beirut in 1915 but, after the outbreak of World War I, a naval blockade cut Beirut off from sea routes. The astronomer therefore accepted an offer to join the Mount Wilson Observatory in southern California. Drawing on his previous experience observing the 1905 eclipse, he assisted in solar research.[112]

Although World War I disrupted scientific cosmopolitanism in parts of Europe and elsewhere, the Syrian Protestant College continued to pursue its transnational mission. 'We are here,' Howard Bliss explained to an Ottoman official, 'not as rivals; we are here to share with the people of the East the best things we have in the West, or rather to exchange the best things that East and West have received. For the whole world needs the whole world.'[113] The SPC indeed depended on its transnational connections and was thus affected by the war. Several American employees of the college returned home after its outbreak. Although the US government did not declare war on the Ottoman Empire, the SPC had difficulties communicating with America and finding replacements. Moreover, many Ottoman subjects who worked for, or studied at, the college were mobilised in the war effort. Adding to that was the breakdown of transportation facilities, food shortages, the partial cessation of remittances from abroad and the depreciation of paper currency.[114]

Even harder hit than the SPC was the Ksara Observatory in the Beqaa Valley, thus exacerbating the war's damage to astronomy in the Levant. The founder of this observatory was Bonaventure Berloty, a French Jesuit who had observed the 1905 eclipse in Spain and was sent to Syria two years later. In the following years, Berloty built up the observatory as an extension of Saint Joseph University in Beirut and as a rival to the Lee Observatory and a similar German institution in Haifa.[115] With French governmental support, he acquired some instruments. However, in 1912, he was still waiting for the 'generosity of a patron' to develop astronomical research.[116] After the Ottoman Empire had entered World War I, it expelled him and other French Jesuits.[117] In 1918, during a time of anarchy between the Ottoman withdrawal and the arrival of the British, local people looted and damaged the observatory. After the war, the Jesuits returned and restored the Ksara Observatory with French government funds. Working closely with the

French army, which was occupying Syria, the Jesuits focused on meteorology and geodesy rather than astronomy.[118]

Transnational astronomical research in Syria thus slowed down during the war years, but did not stop. Following Joy's departure in 1914, Mansur Jurdak, a former student of Raymond Dugan, became acting director of the Lee Observatory. During one summer night in 1918, he was showing visitors Aquila, a constellation of particular interest to many Arabs. Its principal star Altair, derived from *al-nasr al-ṭāʾir*, meaning 'the flying eagle', appeared in well-known poetry. As Jurdak observed the constellation, he noticed a new star that would be classified the nova V603 Aquilae. He sent his discovery via telegraph and mail to Germany. There, they appeared in *Astronomische Nachrichten* together with observations of the same phenomenon from Germany, Austro-Hungary, Denmark and Estonia.[119] Even after the US entered World War I in 1917, an Arab scientist at an American college in the Near East was thus still able to publish in a German journal.

Despite the Ottoman defeat at the hands of America's allies, the end of World War I came with a promotion for former Ottoman subjects such as Jurdak at the Syrian Protestant College. In 1920, as the French occupiers were separating Lebanon from the rest of Syria, the college was renamed the American University of Beirut. This formalised a popular designation.[120] By that time, the institution 'was no longer Syrian, Protestant, or a College', Fuad Sarruf commented.[121] However, while the name change officially Americanised the institution, it was accompanied by the establishment of greater equality among its faculty members. Previously, professorships with full voting rights had been mostly reserved for Americans and Britons, as the New York-based Board of Trustees had sought to keep control of the institution; Arabs were restricted to the ranks of instructors and adjunct professors. In 1920 however, the trustees declared that 'hereafter no distinction shall be made between Anglo-Saxon and non-Anglo-Saxon teachers and administrators in regard to ranks or categories of promotion.' Immediately after this decision Jurdak received full voting rights within the university.[122] Unlike Yakub Sarruf and Faris Nimr, he was thus also able to establish himself permanently at the university. Between 1921 and 1942, he chaired the university's engineering department. At his retirement in 1947, he received the Golden Medal of Merit from the president of Lebanon for educational services to the country.[123]

In addition to his administrative duties, Jurdak continued his astronomical research together with his peers at East Coast universities. While it relied on translational connections, this research was also Arabic in character. One of his most important works was an Arabic and English *Astronomical Dictionary*, which he began in 1900 as a student of Raymond Dugan and published fifty years later. Jurdak had started writing down the Arabic names of the constellations and stars in Charles Young's book *Lessons in Astronomy*. As acting director of the Lee Observatory between 1914 and 1919, Jurdak worked intensively on the dictionary. In 1923, he showed the results to David Eugene Smith, a mathematician at Columbia University, and to Dugan at Princeton University, who encouraged him to complete his work. Smith hoped that Jurdak, with his knowledge of astronomy, Arabic and English, would correct 'errors' in the Western knowledge of Arab astronomy. These errors had 'crept in', as previous Latin translations of Arabic manuscripts were 'neither fair nor very authentic'. Smith added, 'We expect you to undertake the duty of supplying us with direct first-hand information about the work and the achievements of the Arabs in this field.'[124]

Jurdak's *Astronomical Dictionary* was not just an exercise in lexicography, but also an Arab nationalist project. In his Arabic introduction – but not his English preface – he explained that his research was motivated by 'pride in our Arabism, culture and faith'. In order to 'respect its present' and to 'hope for its future', a nation needed a 'solid link to its past'. If the nation is aware of its 'glorious past', it will feel 'greatness', achieve 'success' and 'assume its rightful place among the civilised nations of the world'. For his project, Jurdak studied Arabic works by al-Biruni, Zakariya al-Qazwini as well as books in English, French and German. The result was a 350-page volume that listed Arabic star-names and explained their meaning, transliteration and pronunciation, although he found some names 'corrupted and twisted' to such an extent that he was unable to trace the original Arabic terms. In several cases, he asked Arabic scholars in

Syria and in Egypt for help, but in vain. He thus described his dictionary as 'a beginning'. He added that it 'is an incentive and a challenge for my students and for others in the Arab countries to keep up, continue, and finish in due time.'[125]

Despite Jurdak's passion, Arab nationalism did not dominate astronomy at AUB during the interwar years. With the promotion of Jurdak to a full faculty member in 1920, the university had, in fact, two professors working in astronomy simultaneously. The other one was Julius Brown, a son of one of the university's trustees. A graduate of Dartmouth College, Brown had studied at Oxford before joining the SPC in 1909. In addition to holding the professorship in astronomy, Brown succeeded Jurdak as director of the Lee Observatory in 1919. Like his predecessors, Brown was interested in variable stars[126] and cultivated connections with astronomers based in Europe and America. In 1923, he was elected Fellow of the Royal Astronomical Society of London.[127] A few years later, he spent a sabbatical at the Mount Wilson Observatory in California, where Alfred Joy was working. He also won a grant of $150 from the US National Academy of Sciences for new equipment for the Lee Observatory.[128] Brown's stays in America thus gave a boost to his research on variable stars. Based on photographic plates from the Lee, Mount Wilson and Harvard College observatories,[129] he and two peers examined the magnitudes of Wolf 359. This was one of the stars measured and catalogued by Raymond Dugan's former teacher Max Wolf. Brown and his co-authors concluded that it 'is by far the faintest star, intrinsically, known at present'.[130] As Wolf 359 was also one of the closest stars to Earth, it would later be incorporated into several works of fiction, including *Star Trek.*

While Jurdak and Brown were working at AUB, the Lee Observatory continued to receive state-of-the-art instruments, but their use declined between the late 1930s and 1940s. In 1929, Brown borrowed a spectrohelioscope from the Carnegie Institution of Washington in order to measure solar emissions.[131] In 1936, a portable four-inch Zeiss telescope was donated to the university. Furthermore, a newly-hired British engineer and his Lebanese assistants were able to maintain or restore most of the equipment. They thus saved the equatorial telescope, which was worth around

$150,000.[132] In 1938, Faris Nimr compared these instruments with the ones he had used at the Lee Observatory in the late 1880s. He said that this comparison was like comparing Italian and Ethiopian weapons during the battle of Tembien in 1936.[133] However, during the year of this battle, Julius Brown became dean of the Faculty of Arts and Sciences and thus took on more administrative duties. After Jurdak and Brown retired, courses in astronomy were discontinued in 1947.[134]

Despite this interruption to research at the Lee Observatory, the AUB scientists and their relatives maintained their connections with international organisations and contributed to the consciousness of a world community. After Lebanon had gained independence from France in 1943, Jurdak's daughter Angela, an AUB graduate herself, represented her country at the United Nations Conference on International Organization in San Francisco in 1945. As one of the few women there, but one of the many AUB graduates representing different Arab countries, she was involved in drafting the UN charter. In the late 1940s and 1950s, she also served as Lebanon's consul in New York.[135] In 1954, the National Lebanese Committee for Astronomy, which Jurdak chaired, joined the International Astronomical Union. The union's president Otto Struve encouraged Jurdak to send a small delegation to the union's general assembly in Dublin the following year.[136] Jurdak sent his deputy on the committee Giorgio Contino,[137] as well as a copy of his *Astronomical Dictionary*, which Struve described as 'a great service to science'.[138] Indicating Jurdak's personal importance for this relationship, Lebanon withdrew from the IAU in 1967, three years after his death.[139]

Despite this withdrawal, Lebanon remained part of the United Nations Educational, Scientific and Cultural Organization (UNESCO), which supported the International Astronomical Union. Within this organisation, Lebanon promoted both internationalism and the Arabic language. Also, as the country's representative served Fuad Sarruf, who had edited *Al-Muktataf* until the end of the Egyptian monarchy. After his return to Lebanon in 1952, he became vice-president of the American University of Beirut. In this position, he maintained good relations with the university's American patrons. In 1957, he received a grant from the Rockefeller Foundation to 'study the public relations policies of universities in the United States'.[140] After becoming a member of UNESCO's executive board in 1966, Sarruf contributed to the adoption of Arabic as a working language within the organisation. Between 1972 and 1974, he also served as chairman of its executive board. In recognition of his work on a pan-Arab level, the Academy of the Arabic Language in Cairo and the Iraqi Academy of Sciences both elected him one of their members.[141]

## RESEARCHING ASTRONOMY AND ITS ARAB HISTORY

While Fuad Sarruf was serving as vice-president, connections with East Coast colleges allowed the American University of Beirut to resume research and teaching in astronomy. The successor to Mansur Jurdak and Julius Brown was Owen Gingerich, a student at Harvard University. During the Korean War, he was drafted out of graduate school, but refused military service based on his Mennonite beliefs. Harlow Shapley, his mentor, showed him a letter from AUB seeking an astronomer to take over the Lee Observatory. The Near East College Association in New York, which represented AUB, had gained approval to hire American conscientious objectors as a form of 'alternative service'.[142] Initially appointed as an instructor and director at the Lee Observatory for two years, Gingerich stayed on for another year and was promoted to assistant professor.[143]

Like some of his predecessors, Gingerich worked on both astronomy and the history of astronomy. In 1951, he wrote two magazine articles about a solar eclipse that darkened the eastern United States.[144] In 1955, before leaving for Beirut, he was also part of a Harvard team that observed an eclipse in Ceylon. This observation was another transnational event, bringing together British, French, Dutch, German, Japanese and American astronomers just ten years after the end of World War II. Like David Todd, Gingerich planned to make a movie of the corona, but clouds obscured the sun.[145] In Beirut, with his students he built a model of Jupiter's satellites out of wire and plastic.[146] Inspired by Ted Kennedy, a mathematician and historian of Arabic astronomy at AUB, Gingerich also researched the history of science. In 1958, he published an article on the naming of Uranus and Neptune in the 18th and 19th centuries.[147]

Gingerich's most important achievement during his three years at AUB was the reopening of the Lee Observatory. He found that observatory 'had an astonishingly extensive astronomical library with all the major journals', even though some recent issues had not been unwrapped. In addition to teaching astronomy classes for students, he started a series of 'open nights' for the wider public at the observatory. In 1957, a newly discovered comet created sufficient demand for six extra open nights for 700 people. Later that year, Gingerich invited Camille Chamoun, the Lebanese president, to observe a lunar eclipse. 'We had a very distinguished set of guests plus a guard of 22 policemen!' said Gingerich.[148]

Like Dugan before him, Gingerich left Beirut in order to spend the remainder of his career at an Ivy League institution. In 1958, he returned from AUB in order to continue his graduate studies at Harvard. The interruption of these studies for his service in Lebanon turned out to be a benefit rather than an obstacle in his career. Although he had been 'very annoyed at being drafted out of graduate school', he had also been 'very intimidated' by his doctoral exams. However, after three years of teaching at and directing an observatory, he returned 'full of confidence' and completed his doctorate in 1962. While he was in Beirut, the resources for astronomical research available at Harvard had also received a boost. In 1955, the Smithsonian Institution moved its Astrophysical Observatory from Washington to Cambridge, Massachusetts, in order to affiliate with the Harvard College Observatory. The Smithsonian Observatory had one of the world's fastest computers at the time, which would not have been available to Gingerich previously. The space race between America and the Soviet Union brought further funding for astronomy. After gaining his doctorate, Gingerich worked both at the Smithsonian Astrophysical Observatory and at Harvard, earning a faculty appointment in both astronomy and the history of science.[149]

Circulating between Massachusetts, Lebanon and other countries, Gingerich remained a proponent of internationalism in science. He thus followed his mentor Harlow Shapley, who had donated telescopes to Poland and China and put the 'S' (for 'Scientific') in

UNESCO after World War II.[150] Gingerich served as president of the International Astronomical Union's commission on the history of astronomy between 1970 and 1976.[151] In 1977, he published a letter in *Physics Today* in which he defended a foreign language requirement for American doctoral programs in astronomy. He argued that although most recent publications were in English, astronomers should master another language in order to 'preserve the great tradition of internationalism in science'.[152]

While working in America, Gingerich kept his connections to Lebanon alive. Astronomical research and teaching at AUB thus remained transnational. Just one year after leaving AUB in 1958, he returned in order to carry out an observation with the Lee Observatory's twelve-and-a-half-inch telescope. In a rare occurrence, Venus occulted Regulus during the summer of 1959. Observers at various locations recorded it, and Gingerich contributed to its global knowledge from Beirut. These observations provided information about the atmosphere of Venus before it could be studied from spacecraft.[153] In 1971, Gingerich spent a sabbatical in Beirut with his family, partly to show two of his children where they were born. AUB agreed to pay for his children's tuition at the American Community School in return for his teaching a seminar. This seminar reunited Gingerich with Ted Kennedy and allowed him to meet two of his students: David King, a doctoral candidate at Yale, and George Saliba, an AUB alumnus and doctoral student at Berkeley.[154] Drawing on his experience at Harvard, he taught Kennedy and their students how to use computers for historical research.[155]

Following the seminar of 1971, Gingerich, Kennedy and King began a transnational research project on medieval Islamic astronomy. The history of Arabic science thus remained a focus of AUB and its US-based associates. King argued that 'the history of astronomy and mathematics in the Islamic Middle Ages has not yet been written, and cannot be until the vast collections of manuscripts preserved in such places as Cairo, Istanbul, Damascus, Meshed and the Yemen are surveyed for the first time.' In 1972, the three researchers initiated a project that was funded by the Smithsonian Institution. In collaboration with the American Research Center in Egypt, they explored the Egyptian National Library's vast holdings of

Arabic astronomical and mathematical manuscripts. King stated that there were several thousand of these documents, most of which were 'untouched by modern scholarship'. Following an agreement with the library in 1973, the scholars catalogued the manuscripts and analysed works of particular importance.[156] As part of this project, Gingerich contributed to an edition of a 13th-century Yemeni astronomical handbook as well as an article entitled 'The Astronomy of the Mamluks.'[157]

During the 1970s, Gingerich not only studied past astronomy, but also gave Arabic names to newly described celestial features. Although not an Arab nationalist like Mansur Jurdak, the American astronomer thus recognised past Arab achievements in human history. After Mariner 10 had taken detailed pictures of Mercury in 1974 and 1975, Gingerich became member of a task group within the International Astronomical Union's Working Group for Planetary System Nomenclature. After 'much discussion and no little controversy', this task group decided to name Mercury's craters after 'great human contributors to the arts and humanities'. The group thus sought to honour 'the creative and artistic heroes of mankind'.[158] Based on the task group's recommendation, the IAU in 1976 approved the naming of four craters after Arab poets and writers of the pre-Islamic and early Islamic periods: Amru Al-Qays, Khansa, Abu Nuwas and Al-Jāhiz.[159]

Decades after he had directed the Lee Observatory, Gingerich still retained his connections to the Near East. After the conclusion of his Smithsonian-funded project in 1979, he continued to write about the history of Arabic astronomy and its influence on science in the West. Drawing on the work of Ted Kennedy and his students at AUB, he argued for creativity and innovation in medieval Arabic science. In 1986, he published an article on 'Islamic Astronomy' in the *Scientific American*. Gingerich contended that Copernicus, in formulating a heliocentric model of the universe, 'was simply working out the implications of an astronomy founded by Ptolemy but transformed by the Islamic astronomers'.[160] In 1987, another article by Gingerich argued for the importance of medieval astrolabes in the introduction of Arabic star names into Europe.[161]

While Gingerich continued his research on the history of Arabic astronomy from Massachusetts, the Lee Observatory passed into the hands of a new director who was neither American nor Arab: the Dutch physicist Frans Bruin, who had arrived in Beirut in 1958. Like his predecessors, he developed an interest in the history of astronomy and cultivated connections with American East Coast colleges. He benefited from the proximity to Ted Kennedy as well as a one-year stay at the Department of the History of Science and Medicine at Yale University in 1964. Bruin replicated ancient instruments and built what he called an Arab observatory in his house. He also distributed *Biruni Newsletters* (a pun on the name Bruin).[162] Under his directorship, the Lee Observatory continued to provide calculations for the lunar calendar. Despite being part of a historically Protestant college, the observatory thus supported Islamic rituals.[163]

However, Bruin was the last director of the Lee Observatory. In 1975, the Lebanese Civil War broke out and soon destroyed Lebanon's most important observatories along with much of the capital's cosmopolitanism.[164] Ted Kennedy retired as professor in 1976,[165] and astronomical research came to a halt. The twelve-and-a-half-inch telescope was vandalised and its lens stolen.[166] In 1980, the observatory closed.[167] Bruin taught his last astronomy course in 1979[168] and resigned in 1982.[169] Two years later, in 1984, the university's president, the Princeton alumnus Malcolm Kerr, was assassinated. The same year, Ted Kennedy left Lebanon, ultimately settling in Princeton.[170] In 1988, Bruin fled the country for England.[171] The Jesuits mostly ceased their activities at the Ksara Observatory in the Beqaa Valley in 1979, and its equipment was transferred to the National Council for Scientific Research. A few years later, the observatory was virtually destroyed.[172]

Unlike what happened following previous conflicts, neither observatory was restored after the end of the civil war in 1990. The Ksara Observatory became the property of a winemaking business.[173] The Lee Observatory remained part of AUB, but housed its Faculty Planning and Development Unit. The observatory's aluminium dome was riddled with holes, its wooden floors rotten and the remains of the telescope covered in rust. In 2013, hopes for reopening the observatory for astronomical purposes were low. Mounib El Eid, an AUB physicist, explained that the city had expanded into the area surrounding the university campus and was causing light pollution.[174]

However, as work at the Ksara and Lee observatories stopped, transnational space science continued in America. Using the academic connections between the East Coast and the eastern Mediterranean, an increasing number of Lebanese graduates moved to America. One of the earliest was Philip Hitti, an SPC alumnus who had become one of Malcolm Kerr's professors and teachers at Princeton University. From there, he shaped the establishment of Middle East Studies in the Americas.[175] As research had become difficult in Lebanon during the civil war, many AUB alumni followed Hitti across the Mediterranean and the Atlantic. A pertinent example is George Helou, who graduated in physics in 1975. 'The starry nights of Lebanon's mountains attracted me to the cosmos,' said Helou. Encouraged by one of his instructors,[176] he moved to Cornell University, where he earned a doctorate in astrophysics and radio science in 1980.[177]

With the Lee Observatory closed, Helou continued to pursue an international career rather than returning to Lebanon. After his graduate studies, he spent a year as a research fellow at the Arcetri Observatory in Italy. Subsequently, he joined the California Institute of Technology to work on infrared astronomy. He participated in major international projects, the first of which was the Infrared Astronomical Satellite, which was launched in 1983. A second project was the Infrared Space Observatory, developed by the European Space Agency and NASA.[178] In 1996, Helou joined the team that developed the Spitzer Space Telescope. These telescopes transcended national borders not only through the participation of scientists of different nationalities, but also by being located outside any particular country. 'Above the atmosphere the sky is cold, the galaxies infrared-bright, and the soft glow of the distant universe mesmerising,' Helou said. The space telescopes thus offered a more detailed picture of the evolution of stars and galaxies than telescopes on Earth. From 1999 onwards, Helou led much of this research as executive director of the Infrared Processing and Analysis Center and deputy director of the Spitzer Science Center at Caltech.[179]

Transnational scientists such as Helou received recognition in both America and Lebanon. NASA awarded him an Exceptional Achievement Medal in 2001 and an Exceptional Public Service Medal in 2010.[180] Also in 2010, the Lebanese Academy of Sciences elected Helou as a member. This academy, which was modelled on the French Académie des Sciences, had only been established three years earlier. Its first scientist president was Edgar Choueiri, a Lebanese–American physicist at Princeton.[181] In 2011, the president of the Lebanese Republic awarded Helou a coat of arms. On that occasion, the astrophysicist visited AUB and other educational institutions in the country. 'I took this recognition as an opportunity to talk about astronomy in general, the universe and the excitement of discovery, and especially the role of infrared astronomy,' Helou said.[182]

Helou never returned to work full time in Lebanon, and neither did Edgar Choueiri, who developed plasma propulsion engines for spacecraft. However, another AUB graduate and historian of astronomy named Ahmad Dallal did. After studying engineering in Beirut, he completed a dissertation on the work of the 14th-century Muslim astronomer Sadr al-Shari'ah at Columbia University in 1990. George Saliba served as Dallal's advisor and Ted Kennedy provided further 'valuable suggestions' on the manuscript. In his dissertation, Dallal argued for the existence of 'creativity and originality' in Islamic astronomy during the 14th century, in contrast to previous accounts of 'cultural decline and stagnation' after the 11th century.[183] Subsequently, he worked as a faculty member at Smith College, Yale, Stanford and Georgetown universities, before returning to AUB as provost and professor of history in 2009.[184]

Dallal thus formed part of the wide circulation of people and knowledge between East Coast and Near East colleges. This circulation continued despite damage from the Lebanese Civil War and another military conflict with Israel in 2006. AUB's president between 1998 and 2008 was John Waterbury, an alumnus of Princeton, the American University in Cairo, and Columbia University. His successor was the Egyptologist Peter Dorman. Born to missionaries at the American University hospital in Beirut, Dorman was another alumnus of Amherst College. During Dorman's and Dallal's tenures, Owen Gingerich received an honorary doctorate from AUB in 2011.[185] As one of George Saliba's teachers, Gingerich considered Dallal his 'intellectual grandson'.[186]

Despite conflicts in the Middle East, solar eclipses continued to attract astronomers and contributed to their international circulation. A pertinent example is Shadia Habbal, an alumna from Damascus University who had received a master's degree in physics from AUB in 1973. Like Helou, she moved to America soon thereafter and earned a doctorate from the University of Cincinnati in 1977. She then joined Owen Gingerich at the Harvard–Smithsonian Center for Astrophysics, where she remained until 2004. Using the data collected by two spacecraft, Galileo and the Solar and Heliospheric Observatory (SOHO), Habbal confirmed that a slow solar wind originated from the axes of streamers.[187] She also provided clues that a fast wind emerged from most of the sun rather than just from near its poles. In a 1997 news article in the journal *Science*, peers from other American institutions described these findings as 'revolutionary' and 'a giant step forward'.[188] Her research earned Habbal a chair at the University of Wales in Aberystwyth in 2000, which she held along with her position at the Smithsonian Astrophysical Observatory. In 2005, she moved to the Institute of Astronomy at the University of Hawai'i, but continued to travel the world with a team nicknamed the 'solar wind Sherpas'.[189]

Even while pursuing much of her career in America, Habbal maintained her connections to the Arab region. She earned a Pioneer Award from the Arab Thought Foundation in 2004 and served on the selection committee for the King Faisal International Prize in Science the following year.[190] The United Nations' *Arab Human Development Report 2005*, subtitled *Towards the Rise of Women in the Arab World*, included her biography under the heading 'luminaries in astronomical sciences'.[191] What brought her back to the region were not only Arab foundations, but also clear skies and solar eclipses. Like David Todd a century earlier, she was also a passionate eclipse hunter. Between 1995 and 2010 alone, she led ten expeditions across the globe. Two of them brought her to the eastern Mediterranean, to Syria in 1999 and to Libya in 2006. During the former, she sought to explore the temperature, density and magnetic structure of the corona. During the eclipse over Libya, she found the observing conditions 'terrific'.[192] She was able to describe an unanticipated extended corona with heavy ion abundance.[193]

Habbal's expeditions to Syria and Libya relied on support by the American and host governments. In a presentation about the Syrian expedition, she and her colleagues acknowledged funding by the US National Science Foundation. She further mentioned 'the generous hospitality and support bestowed upon the team by the Syrian Ministry of Higher Education.'[194] The 2006 expedition resulted from a request by the Libyan government to the US State Department that American scientists join their Libyan peers in observing the eclipse. This request came after Libya had abandoned its weapons of mass destruction programs and joined America's war on terror, but before the restoration of full diplomatic relations between the two countries. In what Habbal described as 'enormous assistance',[195] Libyan military planes and helicopters brought the astronomers, their equipment and supplies to the observation site in the southern Sahara Desert. This successful cooperation led NASA and State Department officials to declare that 'science eclipses politics'. In another invocation of cosmopolitanism, they explained that 'science can transcend language and borders, and relatively rare natural events – like a total solar eclipse – provide an opportunity to bring together researchers from different cultures.'[196]

These instances of collaboration raise the question of how important diplomatic relations were for the development of modern astronomy in the eastern Mediterranean. Friendship between the Lebanese and American states, to which AUB itself contributed, were certainly an important factor in the longevity of the Lee Observatory. Precisely how important these relations were, however, can only be assessed in comparison with other cases. Of special interest are Arab countries that also had long-standing connections with American universities, but more volatile – and sometimes disrupted – relations with the USA. Despite vast differences in size, the best comparison with Lebanon might be offered by Egypt. The American University in Cairo, a product of the American Mission in Egypt, opened in 1920.

After this chapter with its focus on Lebanon, the next one pays more attention to Egypt. Like Beirut, Cairo was at times both very cosmopolitan and a hotbed of Arab nationalism. Although both cities were very open to the Arab region and the wider world, Egypt as a non-aligned country was perhaps more flexible during the Cold War. Whereas the Lebanese government largely relied on America and France for support, Egypt also accepted economic and military aid from the Soviet Union. The next chapter will explore the role of space science in American and Soviet cooperation with Egypt. Did this collaboration with both superpowers make Egyptian space science even more global than the Lebanese, whose connections were largely confined to Europe and North America? It is also interesting whether the space programs of the superpowers themselves became more international through the participation of a non-aligned country from the Global South. If so, did this internationalism in space emerge despite or together with Arab nationalism? Finally, the next chapter will explore the scientific networks and circulations underlying this internationalism.

# CHAPTER 4

# Stars of Science

In 2009, Qatar Foundation launched the television show *Stars of Science* (*Nujūm al-ʿulūm*), featuring competitions between young Arab innovators for prizes based on viewers' and jurors' votes. While the show sought to turn the contestants into celebrities, some of the jurors themselves had already achieved this status. They included Farouk El-Baz, an Egyptian-American scientist involved in the Apollo program.[1] In 2001, he already had nicknames such as 'Mister Moon, Mars and Planet Earth'.[2] His work earned him not just honours from NASA and a Pioneer Award from the Arab Thought Foundation, but even recognition in the *Star Trek* universe. In several episodes of *The Next Generation*, a starship named *USS Farouk El-Baz* and a shuttlecraft named *El-Baz* appeared.[3] *Star Trek* and *Stars of Science* helped Farouk El-Baz become a small icon of global popular culture. Indicative of his celebrity, his Facebook page had received around two million 'likes' by 2017. El-Baz's stardom even outshone other prominent Egyptian scientists with NASA such as Essam Heggy, whom around 300,000 people 'followed' on Facebook at the same time.[4]

While few Egyptian space scientists became stars, many could be described as cosmopolitans. The astrophysicist Vasily Fesenkov from the Academy of Sciences in the USSR noted in 1958 that 'as a rule all qualified workers in Egypt in either science or engineering have obtained or are obtaining their training and scientific degrees abroad.' His Egyptian peer Ibrahim Helmi Abdel-Rahman agreed, but had a different take on the situation. He was alarmed that 'most science' in Egypt was 'foreign' and 'the least of it was rooted in the Arab environment.'[5] Nevertheless, cosmopolitan interactions were not just restricted to a scientific elite. Fesenkov had the impression that 'all students' from Cairo University 'are familiar enough with the English language to speak freely in it and to take notes. In addition, every inhabitant of Egypt who wears European clothes speaks English and very often French.' Fesenkov added, 'Among persons, such as the police, who often have to deal with foreign tourists, there are many with a knowledge of many European languages.'[6]

Like the astronomers affiliated with the American University of Beirut, their peers at Egyptian institutions circulated between America and the Middle East. These institutions included not just AUB's counterpart, the American University in Cairo, but to a greater extent Egypt's larger public universities. A pertinent example is Zdeněk Kopal's early co-author Salah Hamid. Hamid received a bachelor's degree from Cairo University in 1944 and a doctorate under Fred Whipple at Harvard University in 1950. Hamid subsequently returned to Egypt, where he worked as an assistant professor at Cairo University. Under Ibrahim Helmi Abdel-Rahman, he later employed his computational skills for economic development at the Institute of National Planning. Hamid maintained a good relationship with Whipple and visited Harvard several times between 1952 and 1962. During this period, the two scientists co-authored articles on the origins of meteor streams for the *Helwan Observatory Bulletin* and the *Astronomical Journal*.[7] In the late 1960s, Hamid rejoined Whipple – as well as Owen Gingerich – at the Smithsonian Astrophysical Observatory in Cambridge, Massachusetts. With funding from NASA, Hamid researched planetary positions.[8] By 1971, the Egyptian scientist had become chief of computer operations at the American University in Cairo. He continued to work on theories of planetary motion with funding from the Smithsonian Foreign Currency Program.[9]

Egyptians such as Farouk El-Baz and Salah Hamid added to the multinational character of the American space programs. These programs did indeed depend on the participation of scientists and engineers from outside the US, including the better-known Germans Wernher von Braun and Johannes Geiss. Whereas von Braun led the development of NASA's Saturn V launch vehicle, Geiss designed physical experiments for the Apollo program with funding from the Swiss National Science Foundation. The public impact of the moon landings was also global, and its rhetoric cosmopolitan. 'We came in peace for all mankind,' Neil Armstrong said. As the astronaut was exploring the lunar surface, Richard Nixon told him, 'For one priceless moment in the history of man, all the people on this earth are truly one.' The first images from the moon were received in Australia and transmitted all over the world, including Arab countries.[10]

Many Arab writers also covered the Apollo program enthusiastically, as they had done with the Soviet space missions. AUB's vice-president Fuad Sarruf, for instance, published a thirty-page article on the 'Conquest of Space' in the Kuwaiti magazine *ʿĀlam al-Fikr*, meaning 'World of Thought', in 1970.[11] Although some of the American rhetoric was part of Cold War propaganda, the Apollo program helped many Arabs envision a cosmopolitan future in space. Watching the moon landing on Lebanese television as a child was a big impetus for Edgar Choueiri to pursue a career in rocket science.[12] In the introduction to his science-fiction novel *Flight into Space*, the Egyptian journalist Husayn Qadri wrote that 'men like us' set their feet on the moon. 'It does not matter, whether their names are John, Peter, Shatalov, Mustafa or Hasan.'[13] In his novel, the three Egyptians selected as astronauts by the United Nations' global space centre are named Farid, Munʿim and Nahla.

Egyptians such as Farouk El-Baz and Salah Hamid made the American space programs even more global than Germans such as von Braun and Geiss did, as they represented a leading country within the Non-Aligned Movement and the Global South. Not formally with or against either NATO or the Warsaw Pact, Egypt, like other African countries during the Cold War, sought cooperation with, and development aid from, East and West. After financing the construction of the Aswan High Dam, the Soviet Union extended millions of dollars in credit for a steel mill in Helwan during the 1960s. Abdel Hamid Samaha, director of the Helwan Observatory, collaborated with the Soviet Academy of Sciences and the Ondřejov Observatory in Czechoslovakia on similar terms as he did with British institutions.[14] Transcending the Cold War division of the world,

Joseph Mikhail, another scientist in Helwan, received funding from the United States Air Force (USAF) for a photometric study of the moon that was published in Kopal's journal *Icarus* in 1968.[15] A few years later, Mikhail used the USAF's Lunar Reference Mosaic for an article that was co-authored with a peer from the Main Astronomical Observatory in Kiev.[16]

What facilitated the circulation of Egyptian scientists was the endurance of institutions that themselves had many transnational connections, similar to the American University in Beirut. The most important Egyptian institution in astronomical research was not the American University of Cairo, however, but the Helwan Observatory. In addition to direct connections with America, it maintained important relationships with British and French institutions. Some Egyptian contributions to American space exploration thus went via Paris and Manchester, where Kopal had established a hub for global astronomical research, especially on the moon and the wider solar system. Having contributed to the American war effort during the 1940s, the Czech–American scientist retained strong links with US institutions from his base in England. In parallel to holding a chair at Manchester, the astronomer was a consultant for the Jet Propulsion Laboratory and a visiting scientist at the Boeing Scientific Research Laboratories between 1961 and 1964.[17] In Egypt, he found astronomers who were just as eager to collaborate across national borders.

## HOT SPOTS

The Helwan Observatory's relationships with British and French institutions went back to the 19th century. Like the American inter-collegiate connections, these relationships were partly sustained by solar eclipse expeditions. Mahmoud al-Falaki presented a report on an eclipse observation in Sudan in 1860 not just to the Khedive of Egypt, but also to the Académie des Sciences. Mahmoud's colleague Ismail Moustapha, also known as 'al-Falaki', observed the same eclipse from Spain as a member of the expedition party of the Paris Observatory.[18] For the August 1905 eclipse, John Reynolds, a Birmingham industrialist, visited Egypt. Impressed by the clarity of the skies, he donated a one-metre telescope to the Khedivial

Observatory in Helwan.[19] In 1909, Harold Knox-Shaw used this telescope to take one of the first photographs that proved the return of Halley's Comet, thus putting Egypt 'on the map of global astronomical research'.[20] Knox-Shaw's subsequent work on southern nebulae contributed to debates about Edwin Hubble's classification of galaxies.[21]

In 1952, the Helwan Observatory used another solar eclipse over Sudan, then a British–Egyptian condominium, for further collaboration with European astronomers. In 1950, Reda Madwar invited Bernard Lyot from the Paris Observatory to mount a joint expedition to Khartoum.[22] In preparation, Khairy Aly, then recently graduated from Cambridge, spent a year in Paris designing two spectrographs together with Lyot. Similarly, Ibrahim Helmi Abdel-Rahman worked for a year at the Leiden Observatory in the Netherlands, where he constructed an instrument for photometry of the corona. Madwar also secured the loan of a camera for coronal photography from the Joint Permanent Eclipse Committee of the Royal Society and the Royal Astronomical Society.[23] In addition, the Helwan astronomers collaborated with their peers from the Royal Observatory in Greenwich, who recorded the partial phase of the eclipse cinematographically from four different locations in Sudan, Iraq and Kuwait.[24]

The eclipse of February 1952, like the one in 1905, provided an occasion for cosmopolitan gatherings. Aly noted that 'from the beginning of January astronomers started arriving at Khartoum from all parts of the world with their technicians and equipment.' The Egyptian astronomer acted as secretary at meetings between the different groups from the physics department of University College Khartoum. These groups included not only Egyptians and Europeans, but also the American Jack Evans from the High Altitude Observatory. Evans worked closely with a team from the United States Naval Research Laboratory. The Jesuit director of the Georgetown College Observatory in Washington concluded the meetings by speaking about the US Air Force observations of the eclipse from Gabon, Sudan and Saudi Arabia.[25]

Despite the new nationalist government after the revolution of 1952, Egyptian astronomers continued to cooperate with their Western colleagues in analysing the results of the eclipse expedition.

A property dispute arose when Bernard Lyot died from a heart attack after returning from Khartoum to Cairo. The Egyptian government allowed the repatriation of Lyot's body, but initially refused to hand over the original photographs of the eclipse to the Paris Observatory. After some negotiations, however, the Egyptian astronomers lent the photographs to their peers.[26] Aly also presented prints of the plates to Hal Zirin from the Harvard College Observatory. Between 1953 and 1955, these plates allowed for the publication of several papers. They appeared in *Comptes rendus hebdomadaires des séances de l'Académie des sciences*, the *Astrophysical Journal* and *Observatory*.[27] Furthermore, the Helwan Observatory collaborated with the Sacramento Peak Observatory in Sunspot, New Mexico, which belonged to the Air Force Cambridge Research Laboratories. Together with Jack Evans and another peer from Sunspot, Aly published a *Spectrographic Atlas of the Solar Corona* as well as two articles on coronal emission lines in 1962 and 1963.[28]

While the astronomers from the Helwan Observatory were disseminating the results of the 1952 eclipse, Grubb Parsons from Newcastle manufactured a new, two-metre telescope for the Egyptian government. Because of pollution from Cairo and heavy industry in Helwan, this telescope was designated for a new station at Kottamia in the desert halfway between the capital and Suez. The access road alone cost £100,000, with another £250,000 required for the telescope and buildings. As part of this multinational collaboration, VÖEST from Linz in Austria, constructed the dome. When advertising the new observatory to their foreign colleagues, Abdel Hamid Samaha and Khairy Aly showed pictures of the construction site as well as sections of the *Spectrographic Atlas of the Solar Corona* at the IAU general assembly in Berkeley in 1961.[29] The Kottamia Observatory started operating as a station of the Helwan Observatory in 1962, and the new telescope saw first light two years later.[30] Also in 1964, Samaha announced at the IAU general assembly in Hamburg that the telescope would be 'available for astronomers from other countries'.[31]

The Egyptian telescope soon generated a sense of competition between astronomers from different nations and even distrust among cosmopolitans. Writing to the American scientific attaché in Cairo in 1964, the Manchester-based Zdeněk Kopal expressed his concern about

Hans Kienle, a German astronomer who had been assessing the Kottamia telescope prior to its acceptance by the Egyptian government. Although Kienle's mission was partly sponsored by UNESCO, Kopal was worried that his German peer might take over the observatory. The Czech–American feared manipulation by the Germans 'as part of their gradual penetration of Egypt of which there is only too much evidence at the present time'. In his view, a German 'caretakership' of the observatory might jeopardise the use of the new telescope for a US-funded lunar mapping program in which he was engaged.[32]

Kopal's concerns did not materialise and the Helwan Observatory, which remained under Samaha's directorship, provided observation time for scientists based in Britain and America almost immediately. In 1964, Richard Woolley, the Astronomer Royal, and Samaha agreed on a cooperation between Greenwich and Helwan. In four expeditions to Kottamia until 1965, Woolley and Aly used the new reflector to obtain the radial velocities of RR Lyrae variable stars.[33] These stars were important indicators of cosmic distance and sources of information on the history of the galaxy. Some of these results made it into in a presidential address by Woolley to the Royal Astronomical Society in 1965.[34]

In 1967, Woolley used the Kottamia station again for research on the halo characteristics of W Virginis variables.[35] He thus furthered the understanding of distinctions between different classes of variable stars. During the same year, Riyadh University invited Woolley to establish an observatory with a large telescope in Saudi Arabia.[36] His work in different parts of the world irrespective of political differences also made the British astronomer a cosmopolitan, although one more concerned with science than morality. In 1972, he became the first director of the South African Astronomical Observatory, where he oversaw the installation of another two-metre Grubb Parsons telescope. He reminded critics of his service to the apartheid government that he had previously worked in Nasser's Egypt, Saudi Arabia and Franco's Spain. Woolley added that he was looking forward to setting up an observatory in hell if the devil would give him a telescope.[37]

In parallel with Richard Woolley's work, scientists associated with Kopal in Manchester and supported by the US Air Force used the Kottamia station for lunar research. These scientists included Jack Saari and Richard Shorthill from the Boeing Scientific Research Laboratories, who in 1963 had published an article on differential temperatures on the lunar surface in Kopal's journal *Icarus*.[38] A total lunar eclipse in December 1964 provided an opportunity to perform further infrared scans. At the suggestion of Kopal[39] and in collaboration with the Air Force Cambridge Research Laboratories, they used the telescope at Kottamia. In several articles published in *Nature*, *Science* and *Icarus*, Saari and Shorthill described 'thermal anomalies' or 'hot spots',[40] sparking a wider discussion about their significance. In an obituary of Saari in 1971, Kopal considered these spots 'probably the most interesting (because unexpected) discovery made on the lunar surface in the years preceding the advent of spacecraft.'[41]

Describing the Kottamia telescope as 'the most powerful astronomical instrument operative on the African continent', Kopal himself was also very keen to collaborate with the Helwan Observatory.[42] Because of its technology, location and transnational connections, Kottamia itself thus became a hot spot for research. In 1956, Kopal had started working with the Pic du Midi Observatory in the Pyrenees, providing his French partners with a one-metre-diameter telescope for work on the moon. After the launch of *Sputnik-1*, he received more than $1 million from the Air Force and NASA to take photographs of the moon. These photographs formed the basis of the Lunar Astronautical Charts, which were to serve NASA in selecting landing sites for the Apollo program. Between 1956 and 1967, the Manchester team obtained over 60,000 photographs at the Pic du Midi, which were processed by the Air Force's Aeronautical Chart and Information Center.[43] Arabic media covered this research as part of their coverage of global science. In 1959, the collaboration between Manchester and the Pic du Midi appeared in an article in the Kuwaiti magazine *al-ʿArabī*.[44]

The American funding in support of the Apollo program allowed for a rapid expansion of the astronomy department at the University of Manchester. At the peak of Kopal's lunar mapping activities between 1964 and 1966, his staff exceeded fifty.[45]

The American funds supported not just lunar mapping, but indirectly other astronomical research. John Meaburn, one of the department's research fellows, helped observe the moon, but mostly pursued other interests. When asked by an Air Force auditor about his lunar work, Meaburn admitted that he was researching interstellar gas clouds. An anguished Kopal quickly added, 'You see, Colonel Downie, John's work puts the Moon into perspective.'[46]

Awash with funds, Kopal and his colleagues did not restrict their ambitions to a collaboration with the Pic du Midi Observatory in France, but reached out to stations further east. To calculate the heights of craters and other features on the lunar surface, the Manchester team sought to take pictures continuously and measure the lengths of shadows with sunlight coming in from different angles. To extend daily observations of the moon from eight to sixteen hours, Kopal thus enlisted an observatory in Japan. From 1964, his team developed a similar collaboration with the Helwan Observatory using the Kottamia telescope. This did not extend the span of the observations in latitude, but increased the number of clear nights for observations.[47] By 1966, the Kottamia station had made more than 2000 exposures at various lunar phase angles.[48] Many photographs allowed for the determination of relative heights on the lunar surface with an accuracy of a few metres.[49]

The global collaboration involving Britain, France, Japan and Egypt succeeded despite animosities in an era of decolonisation. In solidarity with other African countries, Egypt had broken off diplomatic relations with Britain in 1965 over the latter's policy towards Rhodesia.[50] This breakup occurred after much friction over the British presence in Aden.[51] In addition, Egypt had supported the Algerian National Liberation Front during its war against France between 1954 and 1962. Jean Rösch, the Algerian-born director of the Pic du Midi Observatory, thus objected to working directly with the Egyptians. Kopal found himself 'unable' to send his student Joseph Mikhail to work on the Pic du Midi because of Rösch's 'attitude'. For the same reason, Kopal objected to the choice of Rösch as an external examiner of Mikhail's doctoral thesis in 1968. Instead, less than a month after the end of the Prague Spring, Kopal suggested a faculty from both America and Czechoslovakia.[52]

In June 1967, the Six-Day War interrupted lunar photography at Kottamia. Kopal and some of his students were in Egypt at the time and saw the observatory and their quarters 'riddled with bullets'.[53] Their USAF-funded research suffered further from the diplomatic fallout from the war, especially Egypt breaking off relations with America. 'We are very eager to return to Egypt for our astronomical observations,' Kopal wrote to Samaha in December. However, he added that 'there are obvious difficulties in visits to countries where neither the United States nor Great Britain have consular representation.'[54] Furthermore, by that time, NASA's Lunar Orbiters had rendered photography from Earth mostly obsolete for the Apollo program.[55] These satellites were built by Boeing and launched in 1966 and 1967. They photographed the moon using a reference grid based on pictures taken by Kopal and his colleagues.[56] The Lunar Orbiters also contributed to solving the puzzle of Saari and Shorthill's 'hot spots'. These thermal anomalies turned out to represent surface regions with numerous large boulders that absorbed heat during the daytime and radiated it during the night or an eclipse.[57]

While lunar photography was thus interrupted, collaboration in lunar photometry between Helwan and Manchester survived the Arab–Israeli wars of 1967 and 1973. Scientists associated with the Helwan Observatory simultaneously maintained relations with their Soviet peers and thus continued to transcend the global Cold War divisions. During the mid-1960s, one of Kopal's students at Manchester designed a photometer capable of recording light intensity in three wavelength bands simultaneously.[58] This photometer was attached to the Kottamia telescope, where Joseph Mikhail used it to study the colours of the lunar surface. In his USAF-funded article that was published in *Icarus* in 1968, Mikhail confirmed the reddening of the moon with increasing phase.[59] Kopal extended the loan of the photometer through the 1970s, even while Mikhail was working with peers from Kiev.[60] During the late 1970s, the British Science Research Council funded another visit by Mikhail to Manchester, thus supporting further study of the reflectivity of lunar regions.[61]

## GEOLOGY OF THE MOON

While the astronomers associated with the Helwan Observatory engaged in various transnational collaborations, the geologist Farouk El-Baz became another important Egyptian scientist on the American space program. Born in a village in the Nile Delta in 1938, he and his family benefitted from an expansion of state education and increasing opportunities to study abroad. Farouk's father El-Sayed was perhaps the first person from his village to study at al-Azhar[62] and was equally ambitious for his nine children. The Egyptian Revolution in 1952 made education free for all, which allowed Farouk and all of his siblings to go to university. Osama, one of Farouk's elder brothers, studied law at Cairo University and earned a doctorate at Harvard Law School. He subsequently joined the Egyptian Ministry of Foreign Affairs.[63] Safa, one of Farouk's younger sisters, studied medicine at Ain Shams University. Thereafter, she followed her husband Hani Khallaf, another diplomat and friend of Osama's, around the world. She pursued her postgraduate studies at the University of New South Wales in Australia and worked as a doctor in New York and Geneva. A cosmopolitan herself, she worked on several projects for the World Health Organization.[64]

Farouk was equally interested in earning advanced degrees abroad, but his passion lay in rocks. As a boy scout, he had gone on numerous trips to the mountains surrounding Cairo, bringing back and labelling stones. This interest led him to enrol in chemistry and geology at Ain Shams University. Following his graduation, his name appeared on a list of people to be sent to the Soviet Union to complete doctorates. His diplomat brother Osama, who had travelled to the Soviet Union several times, objected, 'You're not going to learn a damn thing from the Soviets. You shouldn't go.' Farouk therefore accepted a scholarship from the Ford Foundation to study in America, starting his first semester in 1960.[65]

In the following years, El-Baz developed strong academic and personal connections to America. Like Kopal, however, he remained itinerant. The US Bureau of Mines assigned him to study geology at the Missouri School of Mines and Metallurgy.[66] While working on his doctorate, he spent a year at the Massachusetts Institute of

Technology. During this time, he met and married Patricia O'Leary,[67] despite her Irish–American family being unhappy with the union. Pat wrote much later, 'My mother decided I hadn't thrown my life away when she began to see him on television and read about him in the press.'[68] After receiving his doctorate in 1964, El-Baz followed his advisor from Missouri to Heidelberg in Germany, where he taught and researched meteorites for a year-and-a-half.[69]

Following his postdoctoral work, El-Baz took his wife back to Egypt with the intention of settling down. After working with geologists in America and Germany, he dreamt of building a school of economic geology in his own country. He shipped four tons of rock samples from Germany and took lifetime subscriptions to journals and memberships in professional societies.[70] He even authored a textbook on the Egyptian Arabic dialect for his wife.[71] However, during the mid-1960s, Egypt's socialist state, which had initially benefited his education, created obstacles for El-Baz's ambitions. While newspapers propagated the slogan, 'the right man in the right place', state rationales conflicted with individual dreams. El-Baz was willing to work as a geologist for any university. However, because of his undergraduate concentration on chemistry, the ministry concerned insisted that he teach that subject at a technical school in Suez. For months, he attempted to take his case to the minister, but failed due to his lack of an 'intermediary' (*wāsiṭah*).[72] Instead of working for the Egyptian state, he thus became an exploration geologist for the Pan-American United Arab Republic Oil Company in 1966. By the end of the year, however, he had decided to return to America as an immigrant.[73]

In America, El-Baz found a permanent job as a geologist, although not as one studying Earth. In early 1967, he applied to universities, but found that many departments had already hired their instructors for the academic year. He therefore reached out to companies working in mining, oil and other fields. In the journal *Physics Today*, he came across an advertisement asking for geologists to work on photographs taken by the Lunar Orbiters. By that time, his wife had already typed more than a hundred application letters. 'Here is one more,' he said. Patricia asked, 'What do you know about the moon?' Farouk responded, 'Nothing, but I can learn.' She typed the letter, and he got the job.[74] His employer, Bellcomm, a subsidiary of American Telephone and Telegraph (AT&T), was created to provide NASA with technical and management advice.[75] AT&T, which was under pressure from antitrust legislation, agreed to lend its expertise not for profit, but as a service to the national quest, 'for the good of our country'. NASA therefore contracted Bellcomm to work for NASA for cost plus $1 per year. Although not formally employed by NASA, El-Baz de facto worked for NASA headquarters.[76]

While NASA offered opportunities for people from various backgrounds, El-Baz still had to earn recognition and responsibilities. He 'spoke funny', looked 'weird', and 'nobody could pronounce' his name, he said.[77] When Bellcomm hired him in March 1967, he straightaway got into an argument with his Welsh supervisor over the conflict between Egypt and Israel. After that, the supervisor concluded that the Egyptian 'probably won't work out'.[78] A few months later, Egypt and other Arab countries broke off diplomatic relations with the US. Egyptian–American relations only formally resumed in 1974 after Nasser's death and another Arab–Israeli war. In order to counteract these combined stigmata, El-Baz had to 'work like hell', he said. He worked through thousands of photographs of the moon's surface that the Lunar Orbiters had taken between 1966 and 1967. For each picture, he wrote the mission number, frame number and surface features on a card. Based on these cards, he compiled a list of sixteen places on the moon that were representative of all surface features and rock types. Quickly recognised as a source of information, El-Baz became secretary of the Landing Site Selection Committee and a spokesperson for geology at NASA.[79]

At Bellcomm and NASA, El-Baz had to balance scientific with political interests. When the preparations for Apollo 11 began, he became supervisor of lunar science planning and operations at Bellcomm. In this way he gained more responsibility for mission planning and oversaw a team of six engineers and scientists. At the same time, he realised that lunar science was not a priority for Apollo 11. NASA engineers repeatedly reminded him that the objective of the mission was 'to land a man on the Moon and bring him safely to the Earth.' While they needed El-Baz's knowledge for selecting a safe landing site, they stressed that the mandate did not include bringing back 'some photos or rocks'. However, El-Baz and his colleagues assumed that 'after Apollo 11, the field would be wide open for science to become the central objective of lunar missions.'[80] Although a foreigner, Farouk El-Baz succumbed to a feeling of national rivalry with the Soviet Union. 'We just wanted to make it to the moon before the Russians; it didn't matter where or how,' he said. He sensed that the mission reflected on America's 'ability as a nation'. A colleague would often say something like 'Guys, listen, we're not going to let the Russkies beat us. What the hell is this? We got to do this, we got to do it right, we got to do it on time.'[81]

In addition to serving as supervisor of lunar science planning, Farouk El-Baz became chairman of the Astronaut Training Group and principal investigator of lunar visual observation and photography.[82] Not just as a foreigner, but also as a geologist, he had to counter stereotypes and earn the respect of the astronauts, who were mostly American test pilots. Dick Gordon, the command-module pilot of Apollo 12, made fun of El-Baz's 'desert origins' and 'camel-riding' expertise. El-Baz also thought that some astronauts 'would not touch

**Figure 4.1** Farouk El-Baz at Mission Control, Houston, Texas, in 1969 (courtesy of Farouk El-Baz)

a geologist with a ten-foot pole' or would say 'when you've seen a rock you've seen them all.'[83]

El-Baz was, however, able to convince the astronauts that they could land their spacecraft more accurately if they memorised certain features of the lunar landscape in photographs. The geologist thus became a resource for the astronauts who were competing with one another for missions. El-Baz's training of Ken Mattingly, who was scheduled to fly on Apollo 13, impressed the latter's colleagues. Stu Roosa, the command-module pilot for Apollo 14, approached El-Baz in the astronaut office, 'Hey, are you Farak El Baez?' El-Baz confirmed this without correcting Roosa's pronunciation. 'I want you to make me as smart as Ken,' said Roosa, before adding, 'Hell no, I want you to make me smarter than Ken!' However, rather than merely imposing his knowledge, he encouraged the astronauts to use their own descriptions of the craters, such as 'snowman' or 'cone', instead of geological terms. Subsequently, El-Baz successfully campaigned for the approval of astronaut-given names by the International Astronomical Union.[84]

In 1970, El-Baz became an American citizen.[85] Although astronauts continued to see him as a foreigner, they increasingly appreciated his background and even spread cosmopolitan messages. They affectionately called him King Farouk, in reference to the Egyptian monarch during whose reign he was born.[86] One of his students was Al Worden, pilot of the command module *Endeavour* during the Apollo 15 mission. After completing his first orbit around the moon, Worden conveyed to NASA's Mission Control Center, 'After the King's training, I feel I've been here before.'[87] In orbit, Worden also read out a message in multiple languages, including Arabic, which El-Baz had taught him: 'mar-HAH-bah al el-ARDE, min-Endeavour ee-LAY-kum sa-lam.'[88]

On seeing Earth from space, Al Worden himself had a vision of the planet as a whole. In 1974, he published a book of poetry with the same title as his Arabic message, *Hello Earth: Greetings from Endeavour*. One of the poems included the following lines:

> Earth: a small, bubbly balloon hanging delicately in the nothingness of space.
>
> Down there we argue of race and ideology and other trivia,
> I am a part of it, and yet apart from it.[89]

When the Apollo program came to an end, another poem by Worden also called for world peace and unity amidst the Vietnam and Cold Wars:

> Say to me, we should be fighting,
> Say to me the world's at war.
> Then I say we are uniting,
> People tired of war, and more.[90]

El-Baz established such good relations with some astronauts that he even authored papers with them. In 1972, he participated with Roosa and Worden at the Third Lunar Science Conference, which was held under NASA auspices in Houston.[91] Together with Roosa, he presented 'significant results from Apollo 14 lunar orbital photography'. Based on 950 photographs taken during the mission, the authors verified the important role of volcanism in the formation of lunar highlands, especially on the far side.[92]

When the Apollo project ended, El-Baz's employer Bellcomm ceased to exist. However, his connections with astronauts helped him continue his space research. When Apollo 17 was announced as the last mission to the Moon in 1972, AT&T prepared to close its non-profit subsidiary. Around the same time, Michael Collins, command-module pilot of Apollo 11, was appointed director of the Smithsonian Institution's National Air and Space Museum. A test pilot but not a scientist, Collins told Stu Roosa at a function, 'They even want me to do scientific research there, and I don't know what the hell they mean by that.' Roosa responded, 'Well, call Farouk.' El-Baz thus left Bellcomm in December 1972 to become research director of the museum's Center for Earth and Planetary Studies.[93] In this position, El-Baz was soon involved in international projects. During the early 1970s, the American government sought to limit the production of strategic nuclear weapons and co-operate with the Soviet Union in space in an effort to reduce tensions.[94] As part of this rapprochement, El-Baz travelled to Moscow in order to negotiate the creation of a joint map of the moon based on American and Soviet pictures. However, the negotiations failed, as the Americans felt that the Soviets were 'hiding things' from them.[95]

During his work for the Apollo program, Farouk El-Baz retained his Egyptian and Arab identities along with a cosmopolitan one. He and his American wife gave their four daughters, who were born between 1963 and 1970, Arabic names: Monira, Soraya, Karima and Fairouz. In 1976, El-Baz said, 'I have not forgotten my link with the Arab world, and I cannot. I came from there. I continually ask myself how I can contribute to scientific development there.' The scientist added, 'I believe one of the best ways I can pay back some of the knowledge I have gained is to use it, particularly for those who need it most.'[96] Part of his service to Arab culture was symbolic. In the 1972 *Apollo 16 Preliminary Science Report,* he recommended naming a lunar basin 'Arabia' to 'honor Arab contributions to astronomy, particularly the study of the Moon'. Some of his suggestions reflected his Egyptian origins specifically as well as his and the IAU's concern for balance between the superpowers. He recommended naming a crater 'Necho' after a pharaoh and 'pioneer geographer', and designating two scarps 'Siberian' and 'Appalachian'.[97]

As a member of the IAU's Task Group for Lunar Nomenclature and contributor to the Working Group for Planetary System Nomenclature, El-Baz gained recognition for most of the Arabic names he recommended, although not 'Arabia'. As he found that the IAU did 'not encourage geographical names', he dropped this designation in favour of 'Al-Khwarizmi', thus honouring an 'Arab scholar' who 'coined the term algebra'.[98] In 1976, the IAU approved this name as well as the names Al-Bakri, Al-Marrakushi, Ibn Battuta, Ibn Firnas and Ibn-Rushd. El-Baz thus honoured further Arab and Muslim writers and scientists. Other approved crater names included those of El-Baz's brother Osama, his mother Zahia, his wife Patricia and their children. El-Baz also named a lunar wrinkle ridge 'Higazy' after a fellow 'Arab earth scientist'.[99] Riad Higazy had been a teacher at Ain Shams University and the subject of an obituary published by El-Baz in 1970.[100]

## DEVELOPING DESERTS

Although a joint map of the moon was not realised, another American–Soviet project in which Farouk El-Baz was involved, succeeded. This project was inspired by the movie *Marooned* starring

Gregory Peck and Gene Hackman. Released after the moon landing in 1969, the film showed Soviet cosmonauts rescuing three US astronauts stranded in Earth orbit. This scenario touched a chord with the Soviets and Americans, who had experienced emergencies during their space programs. Both governments thus agreed on a mutually compatible docking system that would make rescues possible. As part of the Apollo–Soyuz Test Project (ASTP), an American and Soviet spacecraft joined in orbit in 1975. This symbolised a détente and the end of the space race. The ASTP was also the final flight by an Apollo spacecraft and the last manned US space mission before the first shuttle flight in 1981.[101]

In the Arab region, El-Baz advertised the Apollo–Soyuz Test Project as a cosmopolitan mission, sometimes with a reference to religion. He wrote the introduction to a collection of translated NASA reports entitled *What after the Moon?* This collection was published in Cairo in 1975. The scientist described the ASTP as 'a unique kind of cooperation'. He regretted that hostility between America and the Soviet Union had resulted in increased military production, especially of nuclear bombs. There was, however, 'no conflict over the ownership' of space, 'because it belongs to the almighty Creator'. El-Baz hoped that the ASTP would result in 'global cooperation in space exploration projects and collective work for the good of man'. The geologist combined his cosmopolitanism with Arab nationalism. He believed that 'the Arab world should be concerned with catching up with other countries in the field of space research.' He recommended the 'study of Arab deserts via satellites operated by Arab hands and the examination of results by Arab experts.'[102]

Through El-Baz, the Apollo–Soyuz Test Project became directly relevant for the whole Arab region. In 1974, the geologist visited the Gulf as a lecturer from the United States Information Agency. According to the US ambassador to Qatar, El-Baz made an 'excellent impression' and was invited by the Qatari Ministry of Education for another visit in January 1975.[103] During this second trip, El-Baz delivered several lectures at Qatar University and participated in a private visit by the Apollo 15 astronaut James Irwin to Doha. In addition, he discussed the possibility of taking orbital photographs of Qatar as part of the ASTP the following summer.[104] The US ambassador was concerned about Irwin's 'fundamentalist religious orientation'. He hoped that the astronaut would avoid 'what could be interpreted as disparaging comments about Islam'.[105] El-Baz, in contrast, was 'extremely well received' by Qatari officials and the public,[106] whom he addressed as 'my Arab brothers'.[107]

In consultation with Qatari officials, El-Baz made desert research one of the foci of the Apollo–Soyuz Test Project. The ASTP's American director offered El-Baz the chance to work on observations and photography with a budget of $250,000. The geologist accepted and was named principal investigator of an experiment that involved taking pictures of Earth and its arid regions in particular. These regions were easy to observe from space because of the rarity of clouds there. Furthermore, El-Baz found that even the 'best experts' from the US Geological Survey and other organisations had limited knowledge of deserts.[108] This had to do not only with limited accessibility, but also with modern geology's European roots, the only continent without deserts.[109] In order to help the astronauts describe Earth's features in greater detail, El-Baz designed a wheel with more than 100 shades of desert and seawater colours.[110]

Following the successful completion of the Apollo–Soyuz Test Project, El-Baz toured the Arab region again, simultaneously promoting cosmopolitanism and American leadership. In 1976, he accompanied the three American crewmembers of the ASTP, interpreting and lecturing on the value of remote sensing in desert research.[111] An embassy official commented that the group's visit to the United Arab Emirates resulted in 'heightened awareness of positive US contribution to space research'. He added that El-Baz was a 'star' in the Gulf and that the astronauts represented the 'very best in American society'. In a telephone conversation with the delegation, Sheikh Zayed bin Sultan, the UAE president, stated that America, 'which knows the way of goodness', will employ it 'in space and on Earth for the benefit of all'. In a subsequent face-to-face meeting, Sheikh Zayed reiterated America's 'lead in extending the benefits of space scientific research and related technology to benefit all mankind'. The embassy official commented that the astronauts, with El-Baz translating, were 'most effective in stressing this positive potential'.[112]

El-Baz's participation in American scientific diplomacy even reached the Muslim parts of China. In 1978, the United States and the People's Republic of China announced the establishment of diplomatic relations. The same year, the geologist became a member of the first American scientific delegation to visit China's northwestern territories since 1949. Landsat images of deserts inspired enthusiasm among his Chinese counterparts. 'You bring the pictures and interpret the geology,' one of them said to El-Baz, 'and after I learn to study the pictures, I can add the knowledge of geography.'[113] Drawing on his Egyptian background, El-Baz made a special impression on Muslims. One interpreter was eager to learn more about Islam from him. 'He was moved by every word I taught him in Arabic,' the language of the Qur'an, El-Baz said.[114]

The Apollo–Soyuz Test Project and his connections with the Arab region reignited El-Baz's earlier interest in deserts. Studying the project's photographs of Egypt, he found colour zones that did not correspond to the mapped boundaries of geological formations. He therefore organised an expedition to Egypt's Western Desert together with five colleagues from Ain Shams University. In two cars, they drove from Cairo to Alexandria and then west along the Mediterranean coast. Their aim was to work out the different colour zones and to study patterns of sand movements and the resulting landforms. Based on the photographs from space and soil samples, the scientists discovered fertile soil west of Alexandria. Reporting on the expedition in 1977, El-Baz claimed that this was the first time that space photographs had been used to outline the extent of arable land that was reclaimable from the desert. The photographs also revealed shifting sands that were endangering fertile soils on the banks of the Nile further south.[115]

Based on his expertise, celebrity status and family relationships, El-Baz managed to become close to President Anwar el-Sadat. In this way he made connections to the Egyptian government that he had lacked during the mid-1960s. In the meantime, his brother Osama had become a confidant of the president, who had helped lay the groundwork for his visit to Israel in 1977.[116] In 1978, Farouk was appointed scientific advisor to el-Sadat in addition to his work for the National Air and Space Museum. El-Baz's task was to select desert tracts for development, so that the Egyptian population could expand beyond the Nile Valley. In this capacity, he organised a second, larger expedition to the southwest of Egypt in 1978.[117] Together with colleagues from Ain Shams University and the Smithsonian Institution, he investigated the area's development potential based on photographs from space. El-Baz held his advisory position until el-Sadat's assassination in 1981. Ironically and tragically, the assassination was in response to the very Camp David Accords that his brother Osama had drafted.[118]

While serving the Egyptian government, Farouk El-Baz continued to propagate cosmopolitanism. In 1978, he published an article entitled 'Spaceflight and what it does for man' in a journal published by the Egyptian Academy of Scientific Research and Technology. He claimed that 'knowledge about the universe' benefits not just America, but 'man everywhere'. El-Baz claimed that the Apollo–Soyuz Test Project had 'reduced the rivalry between the Eastern and Western blocs'. Moreover, the scientist recommended the use of

**Figure 4.2** Anwar el-Sadat (left) and Farouk El-Baz in 1978 (courtesy of Farouk El-Baz)

'space images instead of unclear maps in teaching geography'. These images 'facilitate the identification of terrains and locations of countries in the whole world'.[119]

Based at the National Air and Space Museum, El-Baz engaged in planetary science beyond Earth and the moon as well. By the early 1970s, El-Baz had already gained IAU approval for naming a valley 'Al-Qahira'.[120] This Arabic word, meaning 'the conqueror', simultaneously referred to Mars and Cairo. The city was supposedly named after Mars, as the planet was in ascendance during its founding by the Fatimids in the 10th century.[121] In the middle of the 1970s, El-Baz compared pictures of Earth taken by the ASTP and Landsat with pictures of Mars taken by the Mariner and Viking spacecraft. The geologist found several correlations between dunes on the blue and red planets. According to him, streaks photographed on Mars had a parallel in sand streaks in the south-western comer of Egypt. He concluded that both appeared to have developed because of the disturbance of sand-carrying wind by topographic highs, 'mountains in the case of the Earth, and craters in the case of Mars'. In 1977, he expected that studying Earth's deserts would thus also further the exploration of Mars.[122] Five years later, he co-edited a NASA volume entitled *Desert Landforms of Southwest Egypt: A Basis for Comparison with Mars.*[123]

In 1982, El-Baz left the National Air and Space Museum, but continued to be involved in studies of Earth from space. By that time, the scientist had reached an annual salary of $54,000, the upper limit for his grade. As his four daughters were all entering college, he was worried about how to pay around $12,000 in tuition fees for each of them. At this point, he received an offer to join the defence contractor Itek Optical Systems as vice-president for science and technology and international development. As El-Baz had previously headed a committee that procured cameras for NASA, Itek's president proposed that he switch sides: 'We'll give you $100,000 as a signing bonus just to say yes.' El-Baz accepted, not just because of the money, but also because of his interest in developing research instruments himself.[124] At Itek, he worked on the Large Format Camera, a high-resolution instrument for the space shuttle.[125] In 1985, however, a serious heart attack forced him to look for a less stressful job.[126]

In 1986, El-Baz thus gave up his position as an industry executive in order to become director of Boston University's Center for Remote Sensing. Under his leadership, this center used radar, multispectral and ultraviolet images for research on Egypt. From a distance, it analysed the tomb of Queen Nefertari, the wife of Rameses II, and a sealed chamber that contained a disassembled 4,600-year-old boat.[127] Introducing space technology to Egyptology, the geologist clashed with more traditional scholars. 'Many archaeologists think I'm nuts,' El-Baz complained, 'They say you have to feel an object, hold it in your hands.' El-Baz, in contrast, was very hostile to what he considered destructive forms of archaeology. He stated that the need to touch an object 'shows total disrespect for what an object is and a great measure of selfishness. I've always thought archaeologists were high-class grave-robbers.'[128]

A 'cosmopolitan patriot',[129] El-Baz expressed loyalty to Egypt in his own archaeological research. 'It was wonderful to take the things I'd learned from the space program – the most sophisticated technology that currently exists – and apply them to studying one of the world's oldest civilisations, which is also my cultural heritage,' he said.[130] El-Baz combined his patriotism with hostility towards previous generations of Western scholars. 'The history of Egypt does not speak kindly of all the people who've gone there and plundered the land of its most magnificent artifacts,' the geologist said in an interview. He then asked, 'What gives you the right to plunder Egypt to fill up your museums in New York and London?'[131]

Besides archaeology, El-Baz continued his desert research from Boston. In 1990, he responded to media reports that blamed East Africans for having caused desertification through deforestation and overgrazing. El-Baz argued instead that local inhabitants had thousands of years of experience of living in drylands. He suggested instead that the global climate, including patterns of air circulation and ocean currents, had formed the deserts. However, the geologist still considered humans capable of developing arid areas. Radar images of Egypt's Western Desert, which were taken from the space shuttle, revealed ancient river courses. Based on these images, the Egyptian General Petroleum Corporation drilled wells and found groundwater. El-Baz estimated that this water was capable of supporting agriculture over 200,000 acres for 200 years.[132]

El-Baz's interest in agriculture was perhaps partly derived from his origins in a village in the Nile Delta. However, it was also derived from a belief, apparently reinforced by his stay in the United States, that a nation's wealth was ultimately derived from the cultivation of land. In 1983, the Qatari magazine *Aldoha* quoted him as saying that 'the United States of America is the only country which has a surplus of food and therefore sits on the throne of cultural progress.' Food security, he argued, gave a society 'peace of mind for the present and the future'. It allowed a society's members to 'think and create' in their free time. El-Baz was therefore alarmed that food production in Egypt was much lower than what was required and that imports were increasing in order to keep up with population growth. He emphasised that the 'country's future security' could only be guaranteed by expanding cultivated land.[133]

El-Baz's belief in agriculture as the basis for an economy perhaps resembles that of 18th-century French physiocrats. However, the Egyptian scientist was also similar to contemporary American visionary engineers – or 'visioneers' – who pursued limitless growth through highly-ambitious plans ranging from space colonies to nanotechnologies.[134] To support an agriculture-based future, El-Baz prepared a plan for the Western Desert: a 'development corridor' reaching from the Mediterranean to the Sudanese border. This corridor would consist of a new 'super highway' and a railroad running parallel to the Nile. It would be connected to water and electricity lines to allow for the establishment of new residential, agricultural and industrial communities. The corridor would have branches connecting it with the main cities along the Nile, including Cairo, Luxor and Aswan.[135] Ultimately, he hoped that this transportation link could be extended to the Sudan and even across the continent to Cape Town.[136] In Egypt, the corridor aimed at preserving agricultural land that contained sediments deposited by the Nile. 'It is a fantastic soil and has survived being used for thousands of years, giving Egypt the best products in agriculture,'

El-Baz said. Ongoing construction, however, led to the continuous loss of agricultural land, which he considered 'very dangerous'. The development corridor would re-focus the construction of future cities in desert areas.[137]

Despite his connections to the Egyptian government, El-Baz found it 'unable or unwilling to pursue the project'.[138] His brother Osama served as an advisor to Hosni Mubarak, but his degree of influence over the president was unclear.[139] By the mid-2000s, his sister Safa had become assistant minister of health and population, and her husband Hani Khallaf a senior diplomat. Privately, the two supported new developments in the desert. Seeking to escape noise and pollution, they were among the first to own a villa in New Cairo, a former wasteland to the east of the capital.[140] In 2005, Farouk presented the idea of the Development Corridor to the government. The following year, a committee of six ministers was formed to review the idea. The Ministry of Planning conducted a feasibility study, but did not present its findings.[141]

With some senior officials interested in his vision, Farouk El-Baz blamed the president for the lack of realisation. In 2012, he described Hosni Mubarak as 'not a technical man and not enthusiastic about science. He never read books himself so he was not a highly intellectual person.' El-Baz added that Mubarak 'would give lip service to some activities but there was no support for science'.[142] Such personal criticism was safer to express after Mubarak had stepped down in 2011. However, the geologist had already complained about the lack of appreciation for science among Arab rulers in an interview for Al Jazeera in 2001. According to him, no Arab 'head of state really understands what science is' and how it benefits industry, development and the people.[143]

After Mubarak's rule, Farouk El-Baz put his hopes in Essam Sharaf, the new prime minister and former minister of transportation, but the politician remained in power for only six months.[144] At the same time, El-Baz's project attracted criticism from his fellow academics. Antar Abou-Korin, an assistant professor of urban and regional planning at Minia University, compared the Development Corridor to Libya's Great Man-Made River and Saudi Arabia's attempt to reach self-sufficiency in wheat. Abou-Korin argued that these giant projects had failed economically and environmentally.[145] Mostafa El-Said, a former professor and minister of the economy under Mubarak, wished El-Baz a long life. However, he argued that in the short term, development efforts should focus on improving education and 'solving the problem of unemployment'.[146]

While advocating the Development Corridor, El-Baz did not depend on Egyptian support alone for his research. Although he was sceptical of Arab rulers anywhere, the scientist continued his relationship with governments in the Gulf. In his 1983 interview for *Aldoha* magazine, he was already advertising the potential of itinerant Arab scientists like himself. 'Why don't we take advantage of these migratory minds,' he asked, 'and seek their help in the field of scientific and applied research in the Arab homeland instead of using foreign expertise?' As an alternative solution to the problem of the 'Arab brain drain', he therefore suggested a transnational circulation of scientists instead of seeing a 'return to the Arab homeland' as the only option.[147]

At the beginning of the 1990s, the Kuwait war created an opportunity whereby El-Baz's own expertise in remote sensing appeared particularly valuable to his audience in the Gulf. In 1991, the scientist gave a presentation on the environmental effects of the war at a conference at Harvard University. In attendance was Ali Al-Shamlan, the director general of the Kuwait Foundation for the Advancement of Sciences (KFAS). A fellow geologist, he understood El-Baz's presentation and agreed to support further research. With KFAS funding, Boston University and the Kuwait Institute for Scientific Research conducted a project on 'Damage Assessment of the Desert and Coastal Environment of Kuwait by Remote Sensing.' The project found that disturbances on the desert surface caused by military activity had resulted in the movement of enormous amounts of soil by the wind, covering new areas with sand. The KFAS found the results so impressive that it decided to fund another project by Boston University and Kuwait University in 1994. This project led to the publication of an *Atlas of the State of Kuwait from Satellite Images*, which El-Baz co-edited. Bilingual in Arabic and English, Al-Shamlan presented it 'to the people of Kuwait and the scientific community worldwide'.[148]

**Figure 4.3** Farouk El-Baz (third from the right) and colleagues meeting with representatives from the Kuwait Foundation for the Advancement of Sciences in Boston in 1992 (courtesy of Farouk El-Baz)

While El-Baz complained about the lack of support for science in Egypt, he helped develop this support in Qatar and America. In 2003, Qatar Foundation and the RAND Corporation established the RAND–Qatar Policy Institute in Doha. The following year, El-Baz was appointed to the institute's Board of Overseers as the only member not affiliated either with QF or RAND. Moza bint Nasser, QF's chairperson, commented that El-Baz's 'efforts will strengthen Qatar's position as a center of world-class scholarship and innovation.'[149] During the same year, the RAND-Qatar Policy Institute developed a proposal for a Qatar National Research Fund (QNRF) as part of QF. After gaining approval, the institute assisted with the establishment of the QNRF between 2006 and 2008. By 2012, the new funding organisation had allocated $500 million for research.[150] In addition, El-Baz gave his name and money for smaller awards for desert research. He contributed the initial endowment for an annual 'Farouk El-Baz Award for Desert Research.' Administered by the Geological

Society for America (GSA), this award subsequently went to researchers based in various countries, including three in Israel by 2009.[151] In addition, Qatar Foundation provided the GSA with $100,000 for the endowment of a second award for desert research named after El-Baz. This annual award, worth $2500, benefitted students and was first given in 2008.[152]

While El-Baz supported projects to create a knowledge-based economy in Qatar, he still remained convinced of the importance of agriculture for the development of African countries. In response to the war that started in Darfur in 2003, he proposed the drilling of wells to alleviate conflicts over water. Satellite images showed a giant ancient lake in the region, promising rich groundwater reserves. In 2007, he travelled to Khartoum in order to brief President Omar al-Bashir and to lecture publicly about the discovery of the lake. The following year, El-Baz visited Darfur to confirm the satellite images. He also met Ban Ki-Moon, the United Nations secretary general, who appreciated the discovery and backed the drilling of wells. Subsequently, El-Baz started an initiative for '1001 Wells for Darfur,' which gained the approval of the Sudanese government. Each well would be named after a donor and was intended to be simple enough so that a child could operate it with a hand pump. Leading by example, students at Boston University had raised $10,000 for a well at one of the most urgent sites by 2013.[153]

While El-Baz's Development Corridor remained a vision, he also continued to promote space science for agricultural development in Egypt. He sought to convince Egyptian leaders to exploit ground-water sources that had been revealed by satellite and radar images. At the same time, he was very critical of people from the older generation. In a 2014 interview, he stated that his country 'has been asleep for some 35 years and its resources have been squandered. The government of Egypt has to do something tangible to develop its resources.'[154] In 2016, President Abdel Fattah el-Sisi asked the scientist to motivate Egyptian young people and strengthen their loyalty to the country. On that occasion, El-Baz told another interviewer that he considered himself part of the 'generation of failure'. 'We have to preserve the post-revolution generation,' he added. 'They can save Egypt, they are the hope. The old ones are finished.'[155] However, before these words could become true, younger scientists had to face conflicts with senior members of the regime, including el-Sisi himself.

## KOFTAGATE

One of the most prominent Egyptian space scientists from the younger generation was Essam Heggy. He shared El-Baz's interest in Martian and Egyptian deserts, but received his postgraduate training in France rather than America. He was born to Egyptian parents in Libya in 1975, after El-Baz had already achieved fame. After receiving a bachelor's degree in astronomy from Cairo University in 1997, he won a scholarship to study at Pierre and Marie Curie University in Paris. In 2002, Heggy completed a doctoral thesis at the Bordeaux Observatory on the use of sounding radar for the detection of groundwater on Mars. He suggested that the techniques involved in his research could be adapted to detect sub-surface water in Earth's dry regions such as Egypt's Western Desert.[156]

Like El-Baz, Heggy also showed himself to be a cosmopolitan traveller as he went between the United States, Europe and the Arab region. After completing his doctorate, Heggy returned to Cairo University's astronomy department, and in 2006 joined the Institut de Physique du Globe de Paris. He adopted French citizenship while keeping his Egyptian one. Furthermore, he worked as a postdoctoral researcher at the Lunar and Planetary Institute in Houston, Texas. In 2009, he said he was 'always travelling', spending 'more than thirty-two hours per month on planes'. The following year, the scientists left the Institut de Physique du Globe de Paris for the Jet Propulsion Laboratory in California. Heggy was unable to keep his position at Cairo University during his transnational career. This was in contrast to previous generations of Egyptian astronomers who had been seconded from their home institutions to teach in Saudi Arabia and other Arab countries. New regulations which limited faculty members' leave to three years forced Heggy to resign.[157]

Similar to El-Baz's work, Heggy's research spanned and transcended the globe. The young Egyptian collaborated with various space agencies and used radar to search for water on Earth, the moon and Mars. Heggy participated in the European Space Agency's Mars Express mission, which was launched in 2003. He also contributed to the Comet Nucleus Sounding Experiment by Radiowave Transmission (CONSERT) experiment as part of the Rosetta mission. This mission also had strong Egyptian associations. The orbiter and its lander were named after the Rosetta Stone and the Philae Obelisk, which had been used to decipher hieroglyphs. In addition, Heggy contributed to miniature synthetic aperture radar experiments on two lunar spacecraft. The first was Chandrayaan-1, which the Indian Space Research Organisation launched in 2008. The second was NASA's Lunar Reconnaissance Orbiter, which left Earth in 2009. At the beginning of 2011, Heggy also used a sounding radar strapped to a helicopter to map Kuwait's aquifers.[158]

Like El-Baz, Heggy became a scientific advisor to the Egyptian presidency. He had long held an interest in developing science and learning in Egypt, having dedicated his doctoral thesis 'to all the children of my country without formal education'.[159] While he was still in his thirties, his work for NASA earned him credibility and respect in his home country. Even Mohamed Morsy tried to benefit from the agency's prestige. During his campaign for the Egyptian presidency, Morsy claimed that he had worked as a consultant on the space shuttle's engines while working at California State University, Northridge during the 1980s.[160] It was after the coup d'état against Morsy in July 2013, however, that Heggy was appointed advisor to the acting president Adly Mansour. Heggy was charged with working on education and water. He was further involved in planning an Egyptian space agency, for which the National Authority for Remote Sensing and Space Sciences submitted a draft law in September. In an interview in January 2014, Heggy stated that the establishment of a space agency would be announced by late March.[161] Heggy himself also became a model for Egyptian pupils, with his biography and image being added to a *National Education* textbook for the first grade of secondary schools.[162]

However, Heggy had difficulty adapting to Egyptian political realities and fell out with the military within less than a year. Comparing his NASA office in Washington to the one in the Egyptian presidential palace, he felt a 'forty-year' gap not only in 'technology', but also in the 'minds'.[163] In February 2014, the Egyptian Armed Forces announced their invention of a 'Complete Cure' device against the human immunodeficiency virus (HIV) and Hepatitis C, which was widespread in Egypt. 'I started working on this project 22 years ago,' said Ibrahim Abdel-Atti, an army doctor, during a press conference. He added, 'It started secretly in the military intelligence department but we are now announcing it to the whole world.' In the audience was Abdel Fattah el-Sisi, then minister of defence and a former director of military intelligence. Heggy, who was on a visit in America, was sceptical and consulted with specialists in Switzerland. Based on their report, he challenged the invention in a newspaper, calling it a 'scientific scandal for Egypt'. 'An issue this sensitive,' he added, 'could hurt the image of the state.'[164]

Abdel-Atti quickly became the subject of ridicule, but the military was still able to regain control over the presidency. At the press conference, the doctor explained how his device worked. 'I take AIDS from the patient and nourish the patient on the AIDS by giving him a skewer of AIDS kofta.' Kofta was a popular meat dish in the Middle East. Soon, hundreds of Twitter users were making fun of this cure, using the hashtag #koftagate. Bassem Youssef, a doctor and comedian who had previously mocked Abdel Fattah el-Sisi, played satirical clips on television. However, with the presidential elections due to take place the following summer at stake, the military attempted to silence this criticism. Abdel-Atti threatened legal action against Youssef in a military court,[165] and pro-military voices accused Heggy of 'treason'.[166] Hoping to avoid an escalation of this controversy, the scientist delayed his return to Egypt.[167] After el-Sisi became president of Egypt in June, he accepted Heggy's resignation.[168]

While Heggy remained in America, members of Egypt's transnational scientific community and political opposition defended him. Nevertheless, his cosmopolitanism and previous work for Western agencies were also used against him. In an interview in the journal *Science*, Islam Hussein from the Massachusetts Institute of Technology complained that few doctors in his home country were speaking out while Heggy was 'attacked by Egyptian media day and night'. Hussein produced scientific videos in order to explain that the Egyptian Armed Forces' claims were 'baseless'.[169] Mohammed Yahia, a journalist with *Nature Middle East*, stated that a paper[170] describing the invention had appeared in a journal published by the World Academy of Science, Engineering and Technology. This was a 'predatory publisher',[171] he explained, which 'publishes pseudoscience and wrong research just to make money out of it'.[172] Mostafa Alnagar, a former member of parliament, asked Heggy in a newspaper article to 'forgive us'. Alnagar remembered accusations on television that the scientist was working for the 'West' or the Muslim Brotherhood and should not be allowed to return to Egypt.[173]

Although Heggy had become a persona non grata for the Egyptian government, he remained a star for many students. In 2015, Heggy took over a professorship of geosciences at Western Michigan University. In October of that year, students at Alexandria University invited Heggy to give a lecture on the evidence of liquid water on Mars. However, because of 'security concerns', the university administration cancelled the lecture one day before it was supposed to take place. An official stated that the university did not have enough time to follow the necessary regulations, including 'checking' Heggy's curriculum vitae and the lecture's topic. The scientist commented on the cancellation by lauding the students' 'passion'. However, he added sarcastically, 'All my regards and appreciation for the university heads and their relentless quest to maintain national security by preventing a lecture on the discovery of Mars.'[174]

Although banned from lecturing in Egypt, Heggy was still able to give interviews to the Egyptian media. In 2015, he explained that his frustrations with the Egyptian military were not limited to the 'kofta device'. He remembered Colonel Ahmed Ali, a military spokesman, announcing in 2013 that the army was honouring three 'astronauts' who had been trained in the Egyptian army. It turned out that the three Egyptians had qualified for the final round of the AXE Apollo Space Academy. This competition, which was sponsored by a brand of men's deodorant, promised winners a future trip on a commercial spaceship. Heggy was critical of the contestants' 'simulated space flight in an amusement park'. He argued that 'astronaut training needs a long time, not two weeks.' Nevertheless, parts of the Egyptian media were hopeful of seeing the 'first Egyptian in space'.[175]

With Heggy in exile and the military in control of the state, the establishment of an Egyptian space agency was also delayed. 'From a scientific, strategic and security standpoint, Egypt needs to establish a space agency,' said Ayman Dessouki, a former chairman of the National Authority for Remote Sensing and Space Sciences in 2015. 'Egypt needs more advanced technology at the current time, and could use a space agency to monitor the country's borders and coastlines through satellite imagery,' he added. Nevertheless, Dessouki explained that the establishment of a space agency required not just a decision by el-Sisi, but also many other procedures and some fifteen years to develop a strategy.[176]

Despite political struggles and challenges in planning, sending an Egyptian into space was, however, not entirely implausible in the 2010s. Saudi Arabia and Syria had already seen two of their nationals fly into orbit during the 1980s. This happened despite both countries having a much shorter history of modern observatories and fewer prominent space scientists than Egypt. These two states were also facing political and financial challenges domestically from oppositionists and drops in oil prices during this decade. The Saudi and Syrian ventures occurred in different contexts. The 1980s saw more intensive Arab cooperation in space technology. In addition, at the end of the Cold War, the American and Soviet governments perhaps had more desire to benefit either politically, strategically, ideologically or commercially from partnerships with Arab countries.

CHAPTER 5

# Space Pioneers

Farouk El-Baz's space research stood in contrast to the cosmological views of an unrelated Saudi scholar with a similar name and fame in the Arab region: Abdulaziz ibn Baz. In 1966, Ibn Baz, then deputy rector of the Islamic University of Medina, published an article entitled 'The Sun moves and Earth is fixed' in Saudi newspapers. In his view, the claim that Earth moved around the Sun contradicted the Qur'an and the Hadith. Hence, it was tantamount to unbelief. The sheikh also stated that God 'spread out' the Earth for his 'worshippers and made it a mattress and cradle for them'. This statement was taken by many to mean that the Earth was flat. This article sparked widespread discussions and drew criticism from other writers. One Egyptian journalist found it 'strange' for these views to be propagated during the 'space age'. To prevent further embarrassment during a time of rivalry with Egypt, King Faisal ordered the destruction of all remaining copies of the article.[1]

Although Ibn Baz was a prominent figure of the Wahhabi scholarly establishment, his cosmological views were at odds with the Saudi government. In general, the Ministry of Information accepted the recent findings of global space science. Moreover, the ministry sought to legitimise them through a connection to the history of Islamic astronomy. In the late 1970s, the minister himself, Mohamed Abdu Yamani, published a book entitled *Scientific Views on the Conquest of Space*. He noted that 'many sons of our Islamic nation' were amazed by the first moon landing and thought that only American and

Russian scientists were 'responsible for this success'. Instead, he argued the moon landing was the result of a long series of experiments and studies by 'researchers of various nationalities and disciplines throughout history'. These researchers included 'our ancestors', Yamani wrote, such as the 10th-century ʿAbd al-Rahman al-Sufi and Ulugh Beg 400 years later.[2]

In Saudi Arabia, interpretations by a religious scholar such as Ibn Baz could co-exist with the views of a Cornell-educated geologist and technocrat such as Yamani. Although both wrote on space in Arabic, they used different sources and rarely referred to one another. Despite the censorship which Yamani's ministry was responsible for, Ibn Baz continued to publish controversial cosmological statements without suffering in his career. The sheikh became director of the Islamic University and chief mufti of Saudi Arabia. After Apollo 11, he expressed scepticism about the moon landing and was again attacked in the Egyptian press.[3] He also republished his earlier article in a book with the following long title: 'The traditional and sensory evidence for the possibility of ascending to the planets, for the motion of the Sun and Moon and for the fixity of Earth.' A second edition of this book appeared with a Riyadh publisher in 1982.[4]

It took the return of a Saudi prince from space for Ibn Baz to cease his assertions.[5] In 1985, Sultan bin Salman flew on the shuttle *Discovery*, becoming the first Arab, Muslim and member of a royal family in space.[6] He was also an employee of the Ministry of Information, and thus somebody who could speak to Ibn Baz with combined royal and technocratic authority. On the occasion of Prince Sultan's mission, many Saudis reconciled scripture with space travel, while also making a pun. They quoted verse 33 from the Qur'anic surah 55 (*al-raḥmān*, meaning 'the merciful'): 'Jinn and mankind, if you can pass beyond the regions of heaven and earth, then do so: you will not pass without Our authority.'[7] The Arabic word for 'authority' used in this verse is *sulṭān*, which happened to be the name of the first Muslim astronaut.

Sultan's space flight was not just important for religious views in Saudi Arabia. The prince represented the entire Arab League, as he oversaw the launch of the second Arab satellite. A Saudi national was chosen for this mission, as the kingdom held the largest share in the

Arab Satellite Communications Organization (Arabsat). However, the flight STS-51-G even transcended the Arab context. It was the first NASA mission that had two non-Americans on board. The other one was Patrick Baudry, the first Frenchman to fly onboard a US spacecraft. A few years earlier, Baudry's colleague Jean-Loup Chrétien had conducted medical experiments on a Soviet spacecraft. In 1985, Sultan joined Baudry in a similar experiment, thus contributing to knowledge that transcended Cold War divisions.[8]

With two foreign astronauts, STS-51-G was an important instance of international cooperation and the global commercialisation of space technology. Under Presidents Jimmy Carter and Ronald Reagan, a series of pro-business and pro-technology laws were passed. During this time, the mentality that 'greed is good' was spreading. After the end of the Apollo program, the US space industry was also trying to become more sustainable and to survive on less government money. NASA thus sought to reduce costs and generate income by launching private and foreign satellites. In order to market its shuttle fleet, the agency even produced a glossy brochure entitled 'We Deliver'. STS-51-G was an important success, as it carried a Mexican as well as an Arab satellite.[9]

In this climate of commercialisation, NASA expanded the circles of astronauts, adding temporary 'payload specialists', which included foreigners, to permanent 'mission specialists'. Offering seats on the shuttle served international scientific cooperation, but was also a marketing tool. 'Launch your satellite on the shuttle and we'll throw in a ride for one of your citizens.' Furthermore, NASA sent members of Congress – who decided about the agency's funding – into space. Finally, for the *Challenger* mission in 1986, the agency selected Christa McAuliffe, a great niece of the Lebanese-American historian Philip Hitti, as the first teacher to go to space.[10] Also part of this mission was Judith Resnik, the first Jewish astronaut.[11]

In accepting foreign payload specialists, NASA competed with the *Interkosmos* program, through which the Soviet Union collaborated with other countries on space missions. As the Arab region included pro-Soviet and pro-American governments during the Cold War, it was able to benefit from both spaceflight providers. Arabic speakers did not differentiate between astronauts and cosmonauts, mostly calling both *ruwwād* (singular *rā'id*) *al-faḍā'*, literally 'space pioneers'. (Less frequent Arabic terms are *mallāḥ al-faḍā'* and *rajul al-faḍā'*, literally 'space navigator' and 'space man'.)[12] In 1985, Mohammed Fares, a fighter pilot from Aleppo, was selected for the Interkosmos space flight program. After many delays, he flew to the space station *Mir* in 1987, thus becoming the first Syrian cosmonaut.[13] After Sultan bin Salman, Fares was the second Arab space flight participant. If one counts Christa McAuliffe as Arab due to her partial Lebanese ancestry, Fares would have been the third. However, McAuliffe did not make it into space, as the *Challenger* exploded less than two minutes into the flight, killing all the crew members. Some people also imagined the Prophet Muhammad as the first astronaut due to his night journey to the heavens. However, Abdulaziz ibn Baz challenged this idea. He argued that Muhammad did not travel through space in the way it is understood in America, the Soviet Union and other countries 'today'.[14]

With Christa McAuliffe and the Prophet Muhammad hardly counting as Arab astronauts, this chapter focuses on Sultan bin Salman and Mohammed Fares. While both benefitted from Cold War rivalry, it would be wrong to see them merely as tools of superpower propaganda to serve 'all mankind'.[15] The global missions of the prince and the 'knight' (the meaning of the name 'Fares') were also the culminations of regional space efforts that went back to the 1950s and were shaped by Arab nationalism and the conflict with Israel. Because of this conflict, Arab states had to struggle in order to access space technology from America in particular. This technology included not only rockets, with their potential military uses, but also entirely civilian satellites. Despite American resistance, many Arabs insisted on owning satellites themselves rather than merely renting transponders on Western ones. While seeking Arab independence in space, however, many of them emphasised the idea of a world community.

## ARTIFICIAL MOONS

From almost immediately after the launch of *Sputnik-1*, Arab scientists were enthralled by satellites. In November 1957, Ibrahim

Helmi Abdel-Rahman published an article on the first 'artificial moon' (*al-qamar al-ṣināʿī*) in the Egyptian magazine *al-Majallah*. As a cosmopolitan, he claimed the Soviet satellite for humanity as a whole amidst Cold War rivalry. He stated, 'The satellite belongs to Earth. It revolves around it from right and left and does not leave it.' In a poetic way, the astrophysicist described that *Sputnik*'s signal was a 'message from the working scientists [*al-ʿulamāʾ al-ʿāmilīn*] to all the worlds [*al-ʿālamīn ajmaʿīn*]'. The signal was also 'a message from the present to the future, a message from the human mind to human civilization, a message of sublimity, superiority and success'. Using words with religious connotations, the Egyptian described the satellite as the 'messenger [*rasūl*] of a new life'. It drives 'the belief [*īmān*] in the unity of humanity' and calls for 'peace and prosperity instead of hunger and war'.[16]

The beginning of the space age coincided with the creation of the United Arab Republic encompassing Egypt and Syria. Therefore, Abdel-Rahman's cosmopolitanism coexisted and at times competed with Arab nationalism in the Arabic press. In May 1959, Satiʿ al-Husri, a prominent nationalist working for the Arab League in Cairo, asked in an article in the Kuwaiti magazine *al-ʿArabī*: 'We were united in the age of camels and mules, how are we divided in the age of the atom and rockets?' Using some of the same terms as Abdel-Rahman, al-Husri described 'Arab nationalism' as the 'belief [*īmān*] in the unity of the Arab nation'. This belief required 'service to this nation' to ensure its 'progress and achievement of greatness, power and perfection in the areas of science, culture, economy, society and politics'.[17]

Eager to appropriate space technology, Arab nationalists soon sought to localise its lexicon. In response to the formation of the United Arab Republic, the Hashemite kingdoms of Iraq and Jordan created a short-lived Arab Federation (al-Ittiḥād al-ʿArabī) in 1958. During the same year, the Iraqi Academy of Sciences began to compile an Arabic glossary on space science. This glossary contained a different Arabic translation of satellite than *al-qamar a-ṣināʿī*, namely *tābiʿ*, literally 'follower'. The list also included a more literal translation of astronaut than *rāʾid al-faḍāʾ*, namely *al-mallāḥ al-falakī*. Some of the terms listed in this glossary did not enter popular usage.

Nevertheless, this and similar works enabled people in the region to think and write about satellites and space exploration in Arabic.[18]

To get ahead of the Soviets, the US soon started offering satellite services to developing countries. A milestone was the Congress's Communications Satellite Act of 1962. It declared a policy 'to establish, in conjunction and in cooperation with other countries, as expeditiously as practicable a commercial communications satellite system, as part of an improved global communications network.' The act explicitly aimed at providing 'services to economically less developed countries and areas as well as those more highly developed'.[19] The United States thus created the Communications Satellite Corporation (Comsat) in 1962 and the International Telecommunications Satellite Consortium (Intelsat) in 1964. While both organisations were designed to benefit the American economy, the preamble to the Intelsat agreement included the desire to promote 'world peace and understanding'.[20] By 1966, fifty-five nations had joined Intelsat. They included Algeria, Egypt, Iraq, Jordan, Kuwait, Lebanon, Libya, Morocco, Saudi Arabia, Sudan, Syria, Tunisia and Yemen.[21]

Intelsat launched its first satellite, nicknamed *Early Bird*, in 1965 and was soon contributing to the consciousness of a global community. In 1965, *Early Bird* beamed a program called *The Town Meeting of the World* across the Atlantic. In June 1967, *Our World*, a more ambitious live television production, included the first performance of the Beatles song 'All You Need Is Love.' *Our World* reached an estimated 500 million viewers in 24 countries. However, shortly after the Six-Day War, the Soviet Union, Poland, Hungary, East Germany and Czechoslovakia withdrew from the broadcast based on their belief in Western support for Israel. One Soviet leader declared that the American, British and West German broadcasting corporations were 'engaged in a slanderous campaign against the Arab countries and the peaceful policy' of socialist states.[22]

The Six-Day War also convinced Arab governments of the need to create a satellite system independent from the West. Shortly after the ceasefire, Arab information ministers who were gathering in Tunisia recommended satellite technology as a means of binding their countries together and countering Israeli propaganda. Two years later, the newly created Arab States Broadcasting Union in Cairo embraced the idea of a regional television network. The union hoped that this network would connect its member states in similar ways as Eurovision and Intervision did in Europe. Initiated by politicians and diplomats, this projected Arab satellite network quickly received support from scientists and technical experts. Missions from UNESCO and the International Telecommunication Union advised Arab governments and recommended the training of specialists in space technology.[23] Farouk El-Baz also saw satellites as important for development in general. In 1975, he presented 'Remote Sensing as a Tool for Development' at the First Islamic Solidarity Conference in Science and Technology in Riyadh. He suggested that, after using American data at first, Islamic countries should purchase or develop satellites 'in order to pave the way for the badly needed economic development'.[24]

Another war with Israel in 1973 and a resulting surge in oil prices gave Arab governments the will and funds to realise the common satellite system. In 1974, Saudi Arabia joined the Arab States Broadcasting Union, but sought to build up a rival organisation with headquarters in its capital rather than Cairo.[25] In 1975 and 1976, telecommunications ministers signed two agreements to form the Arab Satellite Communications Organization. This organisation had an initial capital of $100 million, which increased to $200 million by 1985. Oil-exporting countries led by Saudi Arabia were the biggest shareholders, making participation in the organisation 'very unequal'.[26] However, every member of the Arab League was represented. Even Mauritania, Somalia and the Palestine Liberation Organization (PLO) each held fractions of 1 per cent. Egypt initially owned around 10 per cent, but its membership in Arabsat, and the Arab League as a whole, was suspended in 1979 following its peace treaty with Israel.[27] Arabsat chose Riyadh as the location of its headquarters. Ali Al-Mashat, a communications professor from the University of Baghdad with family roots in Mecca, became the director general.[28]

Amidst regional rivalry and slow progress in the development of the common satellite system, Arab countries continued to rely on Intelsat. In 1969, the British company Cable & Wireless opened in

Bahrain the first terrestrial station for Intelsat in the Middle East and Africa. One engineer described Bahrain, a 'major commerce centre for over 100 years', as the 'logical choice' for the Gulf.[29] However, other Arab countries soon caught up in constructing Earth stations. Even relatively poor Oman, as part of a 'giant leap into the future', inaugurated an Intelsat station in 1975. A colour television program from Washington featured Sultan Qaboos speaking to his ambassador.[30] Two years previously, Algeria had become the first developing country to request a bulk lease of a transponder, proposing a flat rate of $1 million per year. Intelsat accepted this proposal, thus setting a precedent for allotments to other countries. By 1978, Oman, Saudi Arabia and Sudan were also leasing Intelsat transponder capacity for domestic networks.[31]

In competition with Intelsat, Arabsat put forward economic as well as political arguments for the launch of additional, Arab-owned satellites. In remote rural and desert areas, small ground antennae would replace landlines, saving millions of dollars in hardware to achieve full signal coverage.[32] Al-Mashat explained that Arab countries were 'aware of the gross inadequacy of telecommunications facilities between them, both in quality and quantity'. Governments were also conscious that this situation was 'precluding the economic, social and cultural development of the Arab World' and was 'delaying the fulfilment of the Arab-shared aspirations in prosperity and unity'. Al-Mashat promised that his organisation would start 'a new era of Arab cooperation'.[33]

With Arab unity remaining an aspiration, Arab officials at least hoped to decrease dependence on Intelsat and the Western media. In 1984, the Gulf Television Organization (Gulfvision) in Riyadh published a book entitled *Satellites: A New Invasion*. Its author was Mohamed Abdu Yamani, the former Saudi minister of information. He complained that the use of Western-controlled satellite networks 'compel us to adapt to their circumstances and timing, irrespective of the nature of the event'. He added, 'I still remember what happened a few years ago when we tried to transmit the Eid prayer to television stations in Islamic countries. We were told that this was only possible after three hours, as if we – God forbid – set the prayer according to Intelsat's timing.'[34] During the 1984 Summer Olympics in Los

Angeles, people in the region were still upset that telecasts of events in which Arab athletes participated were either ignored or edited severely by producers who were sharing telecasts and satellite time. An Arab satellite system would give its owners the ability to broadcast as much as they wanted of any event in which their athletes were competing.[35]

Independence from foreign media also promises protection from cultural influences. This was especially attractive for the Gulf states, which experienced a religious 'awakening' (*saḥwah*) after the Iranian Revolution in 1979. During this period, national television expanded swiftly, and the Gulf states sought to limit the import of programs produced in the West and elsewhere in the Arab region. This was due to the perception that these productions included morally and politically dangerous content, covered irrelevant issues and did not use Gulf Arabic dialects.[36] 'In order to protect the Arab mind from dangers', the Qatari magazine *Aldoha* reprinted one of the chapters of Yamani's book under the title 'Space Satellites: A New Cultural Invasion'. Yamani warned of satellite broadcasting's 'unchecked influence on other cultures and values, whether religious or social'.[37]

Instead of importing foreign productions, Arabsat promised to enable the Gulf states to *export* their own productions to the rest of the Arab region. Saudi Arabia's largest share in Arabsat was thus also a tool for the Saudi government to gain cultural leadership and soft power in the region. Yamani himself was very aware of space power. Whereas previously 'those who ruled the seas were the ones who ruled the world,' he wrote, 'today it is those who rule space.' Communication satellites were crucial for this power, as he explained in his book *Satellites: A New Invasion.* Those who controlled 'technology' and 'information' were the ones who 'controlled the world'. Yamani also said that 'television is a global media tool in every sense of the word'. It 'crosses global borders in seconds, engaging in an intellectual invasion of every place in the world.' In this context, Yamani mentioned the increase in the number of Earth stations for satellites in the Arab region from five to twenty-one since the beginning of the 1970s.[38]

However, while Arabsat's aims included independence and power vis-à-vis Western countries, the organisation recognised the necessity of using expertise from outside the region to launch its satellites. 'One should be realistic,' Ali Al-Mashat said, 'even in Europe there are only two or three places which can manufacture satellites.' He hoped that by cooperating with foreign institutions, Arabsat would learn from them. He elaborated that 'the objective is to make use of this technology to educate – technically educate – experienced engineers.' Arabsat thus sent its employees for training in Japan, France and America, and invited experts to provide training in Saudi Arabia. The organisation was also eager to have its staff participate in space flights and the launch of its satellites.[39]

Arabsat collaborated with major Western companies in planning its satellite network. After an initial tender, Arabsat commissioned Comsat as consultants in 1977. Following another tender, Arabsat awarded a contract to Aérospatiale for over $134 million for three satellites. This was the first time a French aerospace firm had become the prime contractor for a foreign satellite system. Hidden as subcontractors behind Aérospatiale were Ford Aerospace and Communications, which received $79 million.[40] Ford had previously been under an Arab boycott following American support for Israel during the war of 1973.[41] By awarding the contract to Aérospatiale, and indirectly to Ford, Arabsat thus strengthened and rebuilt links with France and America. To launch the satellites into orbit, Arabsat communicated with the European Space Agency (ESA) and NASA. The organisation also considered the Soviet Union as a partner, but found it less open to sharing technology and allowing access to control rooms and launch sites. In addition, Arab officials feared 'problems' in getting a French- and American-built satellite through Soviet customs.[42]

Arabsat thus relied mostly on European and American technology and expertise to launch its satellites. It chose ESA's Ariane for *Arabsat-1A* and NASA's space shuttle for *Arabsat-1B*, with the third satellite remaining on Earth as a spare. In 1980, the organisation paid ESA $100,000 as the first instalment for the launch of *Arabsat-1A*. After further negotiations, a price of $23 million was set for the launch.[43] In 1982, Arabsat signed a separate agreement with NASA for over $11 million. The reasons for dividing the launch deals were partly diplomatic, in order to strengthen relations with both Europe and America.[44] Another reason was to create competition among the bidders and thus better terms for Arabsat. Finally, the organisation sought to minimise the risks by 'not putting all its eggs in one basket'.[45]

Diversifying the launch vehicles turned out to be prudent. In February 1985, an Ariane rocket transported the first Arab satellite from French Guiana into orbit. One expert described Arabsat as 'a success story in every respect', from the multilateral collaboration to the transfer of technology.[46] The celebrations were short-lived, however. After reaching geostationary orbit, one of the satellite's solar panels failed to extend fully, resulting in reduced performance. Subsequently, the gyroscope which controlled the satellite's orientation malfunctioned. As a result, *Arabsat-1A* was relegated to a backup. In September 1991, another problem appeared in the spacecraft's attitude control system, and it began to drift eastward, failing completely six months later.[47]

## MULTICULTURALISM IN SPACE

The launch of the Arab satellites faced not just technical, but also political obstacles. At the time, the US government considered satellite-related technologies as 'munitions' and therefore as falling under the International Traffic in Arms Regulations.[48] Mohamed Abdu Yamani agreed, describing satellites in his book as 'the new weapons' used by America and the Soviet Union in their 'neo-colonial game'.[49] When Aérospatiale received its contract in 1981, Ronald Reagan's administration met fierce opposition from US senators and Israel to the sale of Airborne Warning and Control Systems (AWACS) to Saudi Arabia. Similar hostility emerged towards the Arabsat deal, which included communications and command subsystems produced by Ford. When the State Department proposed an export license to Congress, it met objections from senators who feared potential military uses of the satellite against Israel. Many legislators also opposed any recognition of the Palestine Liberation Organization or improved relations with Libya. Ford argued that the deal would create jobs and demonstrate American competitiveness vis-à-vis Western Europe. After intense lobbying efforts, an export license approval was granted in 1982.[50]

The same concerns over Israel's security arose in the Senate, when NASA sought to launch the second Arab satellite. Because of this opposition, the agency insisted on removing the names of Palestine, Libya and South Yemen from the contract.[51] This allowed a NASA official to defend the deal, arguing that it 'in no way represents US recognition of the PLO. They are in a number of UN bodies, and we don't pull out of them. This was a straight commercial deal.' In 1985, the agency repeated that the contract represented an agreement with an international organisation, but not its individual members. Hence, the deal did 'not constitute recognition or imply political endorsement by the US of either the PLO or Libya'.[52]

*Arabsat-1B* had a higher profile than its sister satellite not only through debates in Washington, but also through its connection with human space flight. During his talks with NASA officials, Ali Al-Mashat 'got the idea of sending an Arab into space'. In 1984, he prepared a proposal for an Arab to accompany the satellite as a 'payload specialist'. NASA accepted, and the Saudi Ministry of Defense and Aviation nominated a number of Saudi pilots, three of whom made it onto a shortlist: the air force officers Abdulmohsen Al-Bassam and Abd al-Rahman al-Balawi, as well as the civil aviator Prince Sultan bin Salman. While Al-Bassam also came from a well-known central Arabian family, Prince Sultan was a close relative of senior members of the House of Saud. He was a grandson of the first Saudi king, a son of the governor of Riyadh and a nephew of both the reigning King Fahd and the minister of defense and aviation. Sultan held a degree in communications from the University of Denver and worked for Saudi television in the Ministry of Information. He thus had the combined advantage of expertise in aviation and communications and proximity to the Saudi government.[53]

After Arabsat had overcome the political challenges, its pilots faced further medical tests by NASA. 'What does psychiatry want to know exactly?' Sultan asked his interviewer, who responded, 'We are looking for someone who is psychologically and mentally stable. We do not want anyone who might be a burden on the flight or for the crew.' 'You mean you do not want anybody who is crazy,' Sultan said. He then asked whether somebody who was travelling in a rocket with huge engines and giant boosters was not a crazy person.

The psychiatrist laughed[54] and concluded that Sultan was 'saner than most of the rest of the astronauts'.[55] After NASA found him and his colleagues fit, Arabsat made Sultan the main candidate and Abdulmohsen Al-Bassam the backup.[56]

In addition to the NASA interviewers, Sultan had to convince his family. When he first contemplated applying for the mission, his father Salman insisted on getting the 'blessing' of the mother and the permission of King Fahd as the 'father of the family'. Despite her worries about safety, the mother agreed.[57] Fahd had concerns over accusations of nepotism, but permitted his nephew to go through the selection process.[58] In the end, Sultan's father denied that royal status had played a role in the decision. 'Among the dozens who have applied for this mission, Sultan has not been chosen because he is the grandson of Abdulaziz Al-Saud,' Salman responded to a journalist. 'This is an important scientific and technical mission, in which an intermediary [*wāsiṭah*] or titles are of no use,' Salman added. 'Only his capabilities have enabled him to be the first astronaut representing the Arab and Muslim world.'[59]

After beating his Saudi competitors, Sultan, like his fellow French crew member, faced the resentment of some American astronauts. Some cynically labelled STS-51-G 'the frog and prince flight'. This cynicism was partly the result of delays that career astronauts experienced before their first mission or between flights. Between the Apollo–Soyuz Test Project in 1975 and the first shuttle flight in 1981, astronauts were grounded despite years of training and preparation. After 1981, career astronauts had to share their seats on missions with politicians and payload specialists from private companies and foreign nations. In April 1985, Senator Jake Garn became the first member of Congress to fly in space. Some astronauts picked fun at Garn's 'training' by posting a sign-up sheet in the NASA Astronaut Office, asking for volunteers to take an eight-week course to become a senator.[60]

Some career astronauts also had concerns that the liberal space policy of the 1980s would result in insecurity. 'This was an unhealthy environment,' Sultan's fellow crewmember John Fabian said when looking back at the year before the *Challenger* disaster. 'We were taking risks that we shouldn't have been taking. We were shoving

**Figure 5.1** Crew of STS-51-G, with the Saudi and French astronauts on the right of the second row (NASA image in the public domain)

people onto the crews, late in the process, so they were never fully integrated into the operation of the Shuttle, and there was a mentality that we were simply filling another 747 with people and having it take off from Chicago to Los Angeles.' Fabian added, 'This was *not* that kind of vehicle, but that's the way it was being treated at the time.'[61] NASA employees also worried about the Arab satellite, which had failed pre-flight safety reviews. 'The crew recommended that it not be flown,' Fabian said, 'the flight controllers recommended that it not be flown and the Safety Office recommended that it not be flown, but NASA management decided to fly it.'[62] According to Fabian, managers feared political embarrassment and the loss of future contracts if they rejected the Arab satellite. The management's attitude ultimately prompted Fabian to leave NASA in January 1986, shortly before the *Challenger* explosion.[63]

Besides safety concerns, some NASA employees had cultural reservations about allowing Prince Sultan on board. 'They were worried for a while about somebody making a statement for Allah and doing something dumb up there,' said John Creighton, the mission's pilot.[64] NASA thus invited Aramco, which had an office in Houston, to give the American crew members a one-day course in Saudi culture. Dan Brandenstein, who was the commander of STS-51-G, said he wanted to make Sultan and the rest of the crew 'comfortable'. He claimed a 'better understanding' of the French, but found Saudi culture 'a total mystery'.[65] 'Don't tell any harem jokes,' Brandenstein told his crew, 'or any camel jokes.' However, the cultural tension ended quickly, as Sultan had spent much of his adult life in America. At his first meeting with the crew, the prince introduced himself with a joke: 'I've left my camel outside!'[66] Brandenstein even found Sultan to be 'more Americanised or more attuned to the American way' than his French colleague. His crew often made 'a subtle-type joke' that Baudry did not understand and Sultan would explain to him.[67]

Based on his cultural and religious background, Sultan had his own concerns about going into orbit. Before the launch, the prince asked Abdulaziz ibn Baz about praying and fasting, as the *Discovery*'s launch was scheduled during Ramadan. The sheikh's fatwa stated that Sultan should pray five times a day according to a watch set at the time of Cape Canaveral, his place of departure. With reference to the shuttle's speed around the globe, Sultan asked Ibn Baz jokingly, 'the astronauts on the shuttle see the sun rising and setting sixteen times a day. Does that mean that one could complete the Ramadan fasting in two days?' The mufti laughed and allowed Sultan to break his fasting during his training and journey, provided that he fasted after his return.[68]

Humour also accompanied the negotiations about the cultural items that Sultan was allowed to take with him to space. During one of the first crew meetings, the prince joked, 'I want to fly a camel.' Dan Brandenstein misunderstood, 'No, no, it's okay. NASA provides all the cameras. You don't need to bring your own camera.' Sultan responded, 'No, I want to fly a camel, so I can have the fastest camel in the world.'[69] The prince did not bring any camels, but a set of astrolabes to remind the world 'that the Arabs had a leading role in astronomy and space science'. The shuttle mission was thus meant to form a continuation of the older 'Arab and Islamic research'.[70] In addition, Sultan took a copy of the Qur'an on board, the first in outer space.[71]

Not only Arabic astrolabes and the Qur'an made STS-51-G one of the most culturally diverse space missions, but also a few other, more controversial items. John Fabian said that Patrick Baudry 'flew as a Frenchman and enjoyed it'.[72] He did not eat the same food as the rest of the crew, but jugged hare, lobster, crab mousse and chocolate pudding.[73] The French astronaut even caused worries in the press by bringing small bottles of wine.[74] Sultan brought dates from Medina on board, but American journalists were more interested in the flags he took into space. A 'fact sheet' issued by Arabsat[75] and a kit distributed by the Saudi Press Agency contained a map of the Arab region that included the territory, but not the name, of Israel. The kit also had the flags of the 22 members of the Arab League on it, including the PLO. A NASA spokesman said that he was unsure about whether Sultan had a PLO flag in his personal kit, although he confirmed that six Saudi flags were on board the *Discovery*.[76]

Adding to the cosmopolitan character of the mission was the participation of scientists from different countries. NASA tested a retrievable small satellite named *Spartan-1* (for Shuttle Pointed Autonomous Research Tool for Astronomy). *Spartan-1* made x-ray observations of the Perseus cluster of galaxies and measured iron abundances and temperatures in its inner and outer regions.[77] In addition, the Department of Defense included a High Precision Tracking Experiment within the framework of Ronald Reagan's Strategic Defence Initiative or 'Star Wars' program.[78] On the Arab side, a team of scientists from the University of Petroleum and Minerals in Dhahran proposed sixteen possible experiments, five of which received NASA's approval. The first and second consisted of photographing Saudi Arabia from space and observing the new moon marking the end of Ramadan. The third experiment comprised mixing and separating water and oil in microgravity, which promised applications in combating oil spills. A fourth experiment observed the interaction of Earth's ionosphere with gas discharged from the rocket engines. Finally, the Saudi astronaut planned to participate in a French medical experiment about the adaptation of the human body to microgravity.[79] One intended benefit was a better understanding of space adaptation syndrome, which consisted of nausea, dizziness and headaches.[80]

Although NASA agreed to all five Saudi experiments, the second raised concerns and tested the limits of multiculturalism on an American spacecraft. In the STS-51-G press kit, this experiment appeared to be an attempt 'to observe the crescent of the new moon with the unaided eye from orbiter windows as it becomes visible close to the western horizon'. This statement concealed the religious nature of this experiment, as the sighting of new moon marked the end of fasting. Dan Brandenstein was therefore worried about whether Sultan would use his spacecraft as an orbital minaret. 'Knowing that he would be at the centre of a shit-storm,' one NASA employee said, 'Brandenstein confronted the prince and made him agree on the exact wording he would use if he discussed the moon observation on the air-to-ground link, wording devoid of anything religious.'[81]

Dress provided another challenge for multiculturalism in space. When the *Discovery* took off in the summer of 1985, it was at the centre of the Arab world's attention. More than 200 Saudis were at the Kennedy Space Center in Florida for the launch, including the scientists from the University of Petroleum and Minerals and twenty-nine princes.[82] In addition, millions of Arabs watched hours of live telecasting via American satellites.[83] Concerns about cultural norms thus developed in the NASA Office of Public Affairs, but did not reach the shuttle. After Shannon Lucid had been seen in shorts on board, Mike Mullane, one of the capsule communicators (capcoms), received a note requesting that the entire crew wear trousers during press conferences. Mullane interpreted that 'Public Affairs was concerned that the Arab world might find it offensive for one of their princes to be seen hovering with a woman's naked legs prominently displayed next to him.' Mullane threw the note in the bin, refusing to tell an American woman on an American spacecraft to modify her dress 'to accommodate the values of a medieval, repressive society, where women couldn't drive cars, let alone fly Space Shuttles'.[84]

Despite these tensions, STS-51-G succeeded as a cosmopolitan venture. During the first telephone conversation with a non-American head of state from an American spacecraft,[85] Sultan received an Arabic greeting from King Fahd, 'I send you and your colleagues Saudi Arabia's best wishes on this flight that will serve Islam, Muslims and the Arab nation, God willing.' Subsequently, the king sent greetings to Reagan and the American people in English.[86] Sultan himself generated a large audience for these words. This young, tall, dark and handsome jet pilot was creating renewed interest even in America, where space flights had become routine. The American Broadcasting Company (ABC) called the prince the 'real celebrity' of the flight. CBS, *USA Today* and the *New York Times* responded with enthusiasm when Sultan, looking at Earth from space, said that 'the troubles all over the world and not just in the Middle East look very strange as you see the boundaries and the border lines disappearing.'[87] Soon, the signals from *Arabsat-1B* also transcended nations. Unlike what had happened to its sister satellite, its launch was successful. Soon thereafter, the satellite broadcast the evening prayer in Mecca and Medina live, followed by images from the Hajj two months later.[88]

STS-51-G was also multinational in the publication of its scientific results. During a press conference, a NASA official declared it 'one of the most successful' shuttle journeys: 'It was a fantastic flight,' he said, adding that '100 per cent of the objectives were accomplished. I take my hat off to the crew.'[89] Prince Turki Al-Saud, one of Sultan's relatives, analysed Sultan's observations of the ionised gas around the shuttle. This analysis made it into a book chapter published by the American Institute of Aeronautics and Astronautics, as well as into Turki's doctoral dissertation at Stanford University.[90] The results of the phase separation experiment of oil appeared in the University of Petroleum and Minerals' *Arabian Journal for Science and Engineering*.[91] Finally, the medical tests undertaken by Baudry and Sultan led to a conference paper presented in California, as well as to articles in the *Saudi Medical Journal* and various European periodicals.[92]

## ONE PLANET

Having succeeded in the deployment of *Arabsat-1B* and most of his experiments, Sultan bin Salman became a national hero. A royal plane brought him, his backup, and the team from the University of Petroleum and Minerals back to the kingdom. 'I saw large crowds of people waiting, decorations and flowers everywhere,' the prince said about his arrival in Ta'if. King Fahd appointed him a major in the Royal Saudi Air Force[93] and awarded him the Order of King Abdulaziz. Commemorating STS-51-G, the Saudi government issued two stamps showing the shuttle, the satellite, the names of the crew members and the NASA logo.[94] Religious scholars also embraced the prince. Abdulaziz ibn Baz sent telegrams congratulating him, his father Salman and the king. Abdullah Nasseef, the secretary general of the Muslim World League, commented in a newspaper that 'the Muslims in the past had been leading in the fields of science and technology.' He added that Sultan's flight 'presents Muslims with opportunities for progress in space science. The encouragement of science is the mission of Islam, because it is a form of worship.'[95]

Other Arab governments also acclaimed Sultan, turning him into a pan-Arab celebrity. The Kuwaiti ambassador in Washington declared that he 'not only represents Saudi Arabia, but also the Arab and Muslim world. He is one of Kuwait's sons and one of the kingdom's sons at the same time.' Similarly, the Iraqi ambassador stated, 'We are proud to see that the Arab youth has reached such a degree of scientific genius. Prince Sultan represents all Arab youth in this mission, especially as the mission is of a scientific nature and is an expression of the cultural dimension of the Arab nation.' Likewise, the North Yemen ambassador saw the mission as 'new evidence that the Arabs have entered the era of modern technology and science. It is also an image that removes some distorted images of the Arabs projected by the Zionist and hostile media.'[96]

In the cosmopolitan sprit of the mission, the Saudi government honoured the American and French crew members. 'We flew on Saudi Airlines, out of New York into Jeddah, *first class*, on a 747,' Steve Nagel recalled. 'They *really* took care of us.' The American astronaut described the trip as 'more out of this world than going into orbit'. He and his colleagues were 'kind of hopscotched all the way around the country, and everywhere we went, it was just a big spread of food, more than you could ever possibly eat, not enough time to eat it. We were treated like kings and queens.' From Jeddah, the astronauts went to Riyadh, Abha and Dhahran, riding camels, flying falcons and visiting palaces and a farm owned by Sultan's family. Spouses were invited along to places that normally excluded women, such as the Royal Saudi Air Force Officers' Club in Riyadh.[97] 'I must have come home with eight or ten leather briefcases, six or eight Swiss watches, a Turkish carpet,' said John Fabian, 'And the women, the spouses, did just as well.'[98]

The kingdom's law and reputation of gender discrimination worked against its cosmopolitan openness and hospitality, however. Shannon Lucid was a feminist who had long struggled against sexism in her own country in order to become an astronaut. 'She had no love for the Saudis,' said Fabian, 'She particularly objected to Saudi treatment of young women.' Creating further difficulties, Saudi law required women to travel with a male family member as an escort. Because Lucid's husband could not make the trip, the Saudis tried to find a loophole by considering one of her colleagues a guardian or designating her an honorary man. Nevertheless, Lucid initially refused the invitation. When her colleagues arrived in Riyadh, Sultan asked, 'Where's Shannon?' After King Fahd had discovered that Lucid was not coming, he called Ronald Reagan, who called the NASA Administrator. According to Fabian, Lucid 'was on the next 747'. However, she barely spent a day in Saudi Arabia, shook the king's hand and returned home.[99]

In France too, Sultan and his American colleagues encountered both cosmopolitanism and problems regarding the company of spouses. In September 1985, they travelled to France for their induction into the Legion of Honour. The prime minister used the ceremony to appeal for peace in space. 'If there is one domain where international cooperation should exist,' he said, 'it's space, because space is the future and we want it to be a peaceful one for mankind.'[100] From Paris, the astronauts went on to visit other French cities, including Toulouse, the heart of the French aerospace industry.[101] The astronauts travelled alone, John Fabian recalled, 'because the French couldn't make up their mind what they were going to do with the spouses. Patrick Baudry was in the middle of a divorce and he had a Russian girlfriend who was the daughter of a KGB agent!'[102]

From France, Sultan returned to America on another tour as a Saudi, Arab, Muslim and citizen of the world. In an address in Boston, he said that he 'represented' almost a 'billion Muslims'. Farouk El-Baz predicted that the mission would 'cause a revolution in the thinking of the Arab youth'. Sultan himself did not think of himself as 'the personification of the Islamic renaissance', as one paper described him. Nevertheless, he said in a CBS interview that because of his space voyage, Arabs were 'reliving' the Islamic civilisation's past achievements in the sciences, which formed 'the basis for what we see now in the space program'.[103]

To high-school students in Dallas, Sultan said he belonged to Earth as a whole, as he recalled his experience looking down from the shuttle. 'The first day and the second day you notice countries,' he said, 'but then the third day as you go around, you start noticing continents,' the prince continued. On the fifth day, 'you see only the Earth. It becomes … one place, one ball. That's why, when I was over L. A. or wherever, I would call that my home, not just when I was over Saudi Arabia.' Despite being an astronaut, he objected to colonies on Mars or in space on religious grounds. 'It's nonsense,' he said, 'God has given us the best place to live. We don't have to waste time looking for somewhere else and adapt ourselves to it. I think maybe we should use our time to try and maintain it. It really is beautiful.'[104] Similarly, Sultan told one interviewer that his 'happiest moment' was re-entering Earth's atmosphere: 'Whatever distance we travel away from Earth, man always feels that this is his home, not space or anywhere else.'[105]

Sultan also promoted this view of Earth as a whole at other international gatherings. He co-founded the Association of Space Explorers (ASE) and participated in its first 'planetary congress' in Cernay-la-Ville near Paris, France in October 1985. At the opening ceremony, the prince repeated his previous statement: 'The first day or so we all pointed to our countries. The third or fourth day we were pointing to our continents. By the fifth day, we were aware of only one Earth.' This sentiment was shared by the Soviet cosmonaut and co-chair of the congress Alexey Leonov: 'Astronauts and cosmonauts are the handful of people who have had the good fortune to see our Earth from afar and to realise how tiny and fragile it really is.' Leonov added, 'We hope that all the people of the Earth can understand this, and treat the Earth as the home in which they are born, the home in which they live, and the home where their children and grandchildren will live after them.'[106]

After serving on the association's executive committee for four years, Sultan hosted the ASE's Fifth Planetary Congress in Riyadh in 1989 with the theme 'Space for Earth'. In Riyadh, he and his fellow space travellers further developed a global vision of the benefits of space. The congress's general statement promised that space offered 'untold new opportunities to create capabilities of improving humankind's life on Earth'. The participants further argued that international cooperation should 'take into account the needs of developing nations'.[107] In another statement on universal space rescue, the congress emphasised astronauts' and cosmonauts' cosmopolitan identity. With reference to the Outer Space Treaty, they stated that 'astronauts are envoys of mankind.' Continuing the spirit of the Apollo–Soyuz Test Project, they hoped that 'universal space rescue capability would unify the efforts of all space faring nations' and provide assistance to all astronauts in distress.[108]

Enthusiastic about the success of STS-51-G, the Saudi government intended to send more astronauts into orbit. In Dallas, Sultan responded positively to a student's question 'Do you hope to go up again?'[109] The Americans were also open to future collaborative missions. 'We are going to see some very good science come out of this,' said one NASA official about the Saudi participation, 'We look forward to them coming back in the future.'[110] The Saudi government concluded several agreements with NASA regarding the participation of five Saudi astronauts every year or two, with five shuttle missions in total. However, after the *Challenger* disaster, NASA grounded all of its shuttles until 1988.[111] When *Discovery* took off again during that year, it did so with a crew consisting entirely of experienced American astronauts.

By the time shuttle flights resumed, lower oil prices and an administrative and financial crisis within Arabsat was hindering Saudi space initiatives. In December 1985, Arab communications ministers dismissed Ali Al-Mashat as the organisation's director general because of malpractice that had resulted in delays and the failure of *Arabsat-1A*. Al-Mashat had ordered some inferior parts, omitted other parts from the satellite and pocketed the money he had saved. In 1986, he returned to Iraq, joining the Scientific Research Council the following year and contributing to Saddam Hussein's rocket and satellite program. In the meantime, Arabsat was running up deficits and governments were slow to pay their contributions and to construct Earth stations. Many Arab countries preferred to use Intelsat because of its cheaper rates and reliable services. In addition, various regimes were wary of Arabsat's community channel, which was shared across the region, fearing that it might be used for propaganda against them. As a result, Arabsat remained underutilised.[112] A Qatari minister of telecommunications described Arabsat as a 'luxury gadget'. Had the idea been discussed today, he said, 'it would have been rejected.'[113]

Despite the enthusiasm about the first Arab astronaut, Arabsat's difficulties also led to disappointment among nationalists. 'The Arab sky is one, but the Arab Earth is not. It is fragmented,' wrote Moufid Fawzy, an Egyptian television presenter and journalist, in *Aldoha* magazine in 1986. 'Political prohibitions govern' the Arab satellite, he added, and 'fears control it.' These fears even extended to the broadcasting of soccer matches, as commentators might mention 'political taboos' and banners in stadiums might have 'slogans that contaminate thought'. Fawzy thus complained that the satellite's 'most important potential', live broadcasting 'without censorship's scissors', was not being used. The journalist also quoted Hassan bin Talal, the Jordanian crown prince, who said at a forum that Arabsat's real crisis was a 'political crisis and a crisis of policy'. Fawzy explained that it was a crisis of 'Arabism and unity'. He asked, 'How could the satellite broadcast a unified program to the Arab countries, if these countries cannot agree on anything?'[114]

Only after the Iraqi invasion of Kuwait in 1990 did Arabsat expand along with satellite television. Live coverage from Baghdad by the Cable News Network (CNN) made the impact of satellite channels clear. Egypt, which had been readmitted to Arabsat in 1989,[115] launched the Egyptian Satellite Channel in December 1990. It leased an Arabsat transponder to broadcast television programs to its troops in the Gulf in order to counter Iraqi propaganda aimed at Arab soldiers.[116] In the following years, several pan-Arab television companies, including the Middle East Broadcasting Center (MBC) and Orbit, provided business for satellite operators. In 1992, Arabsat launched the spare satellite *Arabsat-1C* using an Ariane rocket. As the demand exceeded its capacity, the organisation purchased 'retired' satellites and renamed them *Arabsat-1D* and *-1E*.[117] In 1996, not only the Al Jazeera Satellite Channel, but also *Arabsat-2A* and *-2B* were launched. A third generation of satellites followed three years later.[118] Al Jazeera and other satellite channels did not bring about Arab unity, but contributed to a stronger sense of pan-Arab identity variously described as 'new Arabism',[119] 'McArabism'[120] and 'everyday Arabism'.[121] By the mid-1990s, however, space scientists from the King Fahd University of Petroleum and Minerals had decided to work directly with NASA without involving Arabsat. For the remote sensing of Arabian geological features, scientists acquired data from the Shuttle Imaging Radar on board *Endeavour* without the participation of Arab astronauts.[122]

Sultan was not involved in Arabsat's expansion, but he contributed to opening up Saudi Arabia more to the world. Following his retirement as a colonel in the Royal Saudi Air Force in 1996, he earned a master's degree in social and political science from Syracuse University. In 2000, he became secretary general of the Saudi Commission for Tourism and Antiquities and, in 2008, its president.[123] Although this organisation lacked the power of other agencies headed by princes, he sought to rebrand Saudi Arabia along multicultural lines. The Roads of Arabia exhibition presented the kingdom's pre-Islamic antiquities in various European and American cities. Sultan commented that this exhibition showed that 'from early times Saudi Arabia has always played an important role in the Middle East. It has existed as a bridge between the East and the West.' The prince continued, 'The first dialogue of civilization finds its origins in the Kingdom, which has historically had a strong communication with the rest of the world.'[124]

In parallel to his work for the Saudi Commission for Tourism and Antiquities, Sultan continued to promote international collaboration and cosmopolitanism in space science. In 2010, marking the twenty-fifth anniversary of STS-51-G, he participated in the Saudi International Space and Aeronautics Technology Conference in Riyadh, which the King Abdulaziz City for Science and Technology had organised in co-operation with NASA. In his opening speech, NASA Administrator Charles Bolden stated that 'STS-51-G launched a valuable satellite in the field of telecommunications: Arabsat.' He added that 'the shuttle program has helped to bind our world closely together through the telecommunications networks that it established.' Sultan expressed a similarly global vision: 'We are today, no matter how our geographical locations, languages, and religions differ, still human beings created by one God, and we live on one planet, and one fate binds us together on this planet.'[125]

In 2011, satellite television, and the pan-Arab identity it strengthened, played an important role in the rapid spread of uprisings across the region. At the same time, Sultan continued to say he belonged to Earth as a whole. Also in 2011, he published a memoir entitled *One Planet: The Story of the First Arab Mission to Space*. This book elaborated on his memory of seeing Earth from space and formed another call for cosmopolitanism. 'You are transformed by this magical view', he wrote. 'A special feeling emerges; you lose the urge to find boundaries between states and countries and the world becomes a globe of seven continents and seas.' Sultan continued, 'The world seems to meld into a large kaleidoscope of color and beauty. No longer do you think in terms of my city or my country, rather you begin to sense that you are a part of something much larger.' The prince concluded, 'The insight l gained in space confirms my deep conviction that we are all part of a much larger whole, one that demands we stalwartly work toward understanding relationships between individuals and nations, compelling us to live and work together toward making our planet a better place.'[126]

While Sultan was publishing his hopeful account, the Arab uprisings plunged Syria into civil war. Some state borders were losing their effectiveness as groups such as the Islamic State claimed territory in multiple countries and millions of refugees fled the violence. The war destroyed much of Syria's infrastructure, including scientific institutions. However, an alliance between Damascus and Moscow that had begun during the Cold War endured. Although this alliance was responsible for a lot of destruction, on at least one occasion during the late 1980s, Syrian and Russian fighter pilots – as cosmonauts – had flown over Syria on a mission of peace.

## COSMONAUTS' COSMOPOLITANISM

Syrian interest in space science became institutionalised in the early 1980s as part of wider efforts in the region to use remote sensing for economic development. In 1981, the government established a National Remote Sensing Center through an ad hoc steering committee. In 1986, a legislative decree replaced this center with the General Organization of Remote Sensing (GORS) located west of Damascus.[127] During the same period, Tunisia also invested in what it called in French *télédétection*. Starting in 1975, the country cooperated with France in utilising satellite images to study water resources, land use and erosion. In 1988, Tunisia founded its own National Center for Remote Sensing and two years later began to host the Centre Régional de Télédétection des Etats de l'Afrique du Nord (CRTEAN). Algeria, Libya, Mauritania, Morocco and Tunisia were the founding members, with Egypt and Sudan joining later.[128]

While Tunisia confined its efforts to North Africa, Syria attempted to integrate the wider Arab region into space science. Ruled by the Arab Socialist Baath Party since 1963, Syria led attempts to teach the natural sciences in Arabic. In 1976, a presidential decree also established the Institute for the History of Arabic Science at the University of Aleppo. This institute partly sought to provide contemporary Arab scientists with inspiration from history.[129] During the Lebanese Civil War, the institute invited Ted Kennedy from the American University of Beirut for a stay. At the institute, Kennedy set up the *Journal for the History of Arabic Science*.[130] In 1990, the General Organization of Remote Sensing organised the First Regional Symposium on the Role of Remote Sensing in Supporting the Economy of Developing Countries. This symposium was followed by the First Arab Conference for Remote Sensing and Space Research in 1995.[131] As early as 1992, GORS was also 'seeking to establish an Arab Space Agency'.[132] The Syrian government was also more ambitious than its Tunisian counterpart in combining remote sensing with human spaceflight.

While NASA's space shuttles were grounded between 1986 and 1988, Soviet spacecraft remained open for scientists from Arab and other countries. Between 1978 and 1984, cosmonauts from Eastern

Europe, Vietnam, Cuba, Mongolia, France and India all flew alongside Soviets on Soyuz spacecraft. Syria, which had provided the Soviet Union with a naval facility since the early 1970s, agreed with its ally that one of its citizens would be next.[133] Initially, dozens of pilots were nominated for the mission, from whom four finalists were chosen: two Alawis, a Druze and one Sunni – all men. Although the Syrian government had sent a delegation to the Soviet Union to 'help' with the process,[134] it ceded the final choice to the Soviets, because, as one general explained, 'they have more accurate criteria for appraisal.'[135]

After a series of tests in Star City near Moscow, the most senior candidate, an Alawi colonel, was removed because of a medical problem, while the Druze was deemed not good enough. Mohammed Fares – the Sunni – subsequently became the primary cosmonaut and Munir Habib, an Alawi from the coastal city of Jableh, the backup.[136] Both had graduated from the Institute of Aeronautics in Aleppo in the early 1970s and had reached the rank of lieutenant colonel in the Syrian Arab Air Force.[137] Fares was initially scheduled to travel to *Salyut 7* in 1985, but was delayed due to the space station's problems, including a loss of power. Eventually, he was assigned to fly as the first non-Soviet citizen on the *Soyuz TM*-3 spacecraft to the space station *Mir* in 1987. This visit was scheduled for July in order to have favourable conditions for taking pictures of Syria from space.[138]

Like STS-51-G, Fares's mission was marked by Cold War rivalry as well as international cooperation and cosmopolitanism. *Mir* means 'peace' and has the homonym 'world'. However, in response to Reagan's Strategic Defense Initiative, the space station was meant to include *Polyus*, an orbital weapons platform. The Soviet Union postponed the launch of *Polyus* as a separate spacecraft in order not to jeopardise negotiations with America. However, the government in Moscow kept the platform until 1997.[139] During a press conference in Damascus in 1986, Fares and Habib asserted that the Soviet–Syrian space flight 'crowns the growing relations between the two countries in various fields'. They added that the mission 'indicates that Syria and the USSR have a joint stand in one trench to confront the dangers posed by the aggressive acts of imperialism and world Zionism'.[140] The Soviet leader Mikhail Gorbachev described Soyuz TM-3 as 'a striking page in the annals of development and strengthening of Soviet–Syrian friendship'.[141] Present at the launch were the Syrian defense minister Moustafa Tlass, an Interkosmos delegate and various Soviet officials.[142]

Although the Soviet–Syrian mission took place during a period of glasnost and greater transparency in Soviet governmental institutions, some mistrust remained. In contrast to many Soviet cosmonauts who had spent months in space, Fares, like other foreigners, was scheduled to stay on board for little more than a week. Similar to payload specialists on American spacecraft, he was also barred from navigation and other sensitive operations. Before the launch, Fares's wife, who was visiting him, suffered an electric shock in the shower. The Soviet crew commander accused the backup cosmonaut Habib, but Fares responded: 'We compete for a trip into space, but we are not killing each other.' Syrian officials were also suspicious. Some of them wondered why, and resented that, the Soviets had chosen a Sunni over an Alawi after years of conflict between the Alawi-dominated government and the Muslim Brotherhood. Fares responded that he was a 'son of this country' like 'every other Syrian' and all of his 'ancestors had been buried within Syrian territory'. Nevertheless, the military attaché informed Fares that a toxic substance addressed to one of the cosmonauts had been found at Damascus International Airport.[143]

Fares was not involved in the satellite launch, but like Sultan bin Salman, he conducted physical and medical research. Accommodating patriotism on an international mission, he brought with him soil from Damascus[144] and undertook experiments named after Syrian historical sites. All of these experiments were prepared by the Syrian General Organization of Remote Sensing in coordination with the Soviets.[145] The experiment 'Qasyun' smelt an aluminium-nickel alloy. Drawing on Arab astronomical heritage, it was named after a mountain near Damascus that was site of an observatory established under the Abbasid caliph al-Ma'mun. 'Bosra' collected data on the upper layers of the atmosphere, and 'Apamea' crystallised gallium antimonide. 'Palmyra' created a crystalline structure resembling human dental and bone tissue as a first step towards producing this material in space.[146]

Like the Apollo–Soyuz Test Project and STS-51–G, Soyuz TM-3 also involved a remote sensing experiment. Named 'Euphrates', it searched for water and mineral resources, studied plate tectonics and assessed pollution over big cities. Fares explained that the latter experiment was the most interesting to him, 'because it tackles some vital issues of concern to my homeland'.[147] On the day of this experiment, 25 July 1987, the Soviets connected Fares and President Hafez al-Assad via telephone and television. During the conversation, the cosmonaut said that he was passing 'for the first time above the sky of our beloved country Syria'. He described it as 'wonderful and beautiful, as it really is', with its 'wonderful coasts' and its 'beautiful green mountains'.[148] Thirty years later, he still said, 'It would take me days to describe the feeling of going up there, seeing Planet Earth, seeing Syria from above, and that sense of pride of accomplishing something historic for my country and for the Arabs.'[149]

Despite patriotism and mistrust, Mohammed Fares sensed and expressed cosmopolitanism and solidarity among space explorers. 'We were like brothers. All the cosmonauts and astronauts know each other and are one big family,' he said. 'Those who have already been in space gave us advice and gave us courage to push on whenever things got hard.' Shortly before the launch, there was 'fear of never seeing your family again, fear of failing in the mission and the fear of disappointing a nation'. However, once in space, he felt 'peaceful'. The space pioneer recalled that *Mir* 'was a home for all the astronauts and cosmonauts that visited it and lived on it, each leaving something behind, from books to sketches to photographs to flags, to all sorts of symbolic items from their home countries. It was cluttered with memories.'[150] In December 1987, Fares and his wife named their new-born son Mir.[151] In 1988, he contributed the following sentence to a book entitled *The Home Planet* by the Association of Space Explorers: 'From space, I saw Earth – indescribably beautiful with the scars of national boundaries gone.'[152] Fares supported this cosmopolitan association further by participating in its twelfth and fourteenth planetary congresses in Montreal in 1996 and in Brussels in 1998.

Fares thus gained global fame and added a cosmopolitan identity to his Syrian and Arab ones. Nevertheless, he soon felt marginalised within his home country. He received the Gold Star Medal of the Hero of the Soviet Union, the Order of Friendship of Peoples and the Order of Lenin, a decoration previously bestowed upon Gamal Abdel Nasser. He also became an honorary citizen of the town of Arkalyk in Kazakhstan. Thousands of Syrians greeted the cosmonaut on this return. The Syrian government made him a Hero of the Syrian Arab Republic, awarded him the Order of Military Glory and named streets, schools and even an airport after him.[153] However, Fares soon found himself confined to his home without work or promotion for long periods. Hafez al-Assad's government forbade him from appearing in the media and at many public events. He explained this marginalisation as resentment within the regime over his selection as the primary cosmonaut over his Alawi comrade.[154] After nine years, the Syrian government made him director of his alma mater, the Institute of Aeronautics in Aleppo, a position with 'empty powers'.[155] He found that 60 to 70 per cent of his cadets were Alawis and that unqualified ones had been accepted 'under pressure' and through 'nepotism' (*wāsiṭah*).[156]

After Bashar al-Assad became president of Syria in 2000, Fares initially gained more freedom to give interviews. In 2003, he participated in the Third Islamic Astronomical Conference in Amman, where he spoke to the president of the Qatif Astronomy Society. In 2007 and 2009, he gave extended interviews to the Syrian and Jordanian newspapers *al-Thawrah* and *al-Dustūr*. In these interviews, he also expressed a strong attachment to Earth as a whole. In 2003, he said that 'the space explorer subconsciously resorts to Earth, when he senses any danger. It is like the mother's lap, to which the child turns and from which it gains kindness and compassion.'[157] Similarly, in 2007, he remembered, 'When I landed on Earth, I did not think in which country I was. What was important to me was that I had arrived on Earth, as if it was the caring mother.'[158]

Although he was a trained MiG fighter pilot, Fares combined his cosmopolitanism with pacifism in some of his interviews. In 2003, he stated that the space pioneer finds the 'occurrence of wars between human beings strange', as 'he feels more than anyone else how this small planet in this big universe is the only one with plentiful sources of life.' The space explorer thus also 'feels the importance of preserving it'.[159] In 2007, he also 'wondered why there is all this evil in the world. Earth is beautiful and can accommodate all human beings.' 'Unfortunately,' Fares added, 'greed and obsession are the evils that control it.'[160]

During the Syrian Civil War, he abstained from violence and – like Moustafa Tlass and other members of the regime – left the country. By that time, Fares had reached the rank of major general (*liwāʾ*) and was head of the Air Force's Department of Organization and Management. At the beginning of the uprising in 2011, however, Bashar al-Assad discharged him and other non-Alawi officers.[161] He tried to leave the country, but was forbidden from travelling and did not possess a passport. In summer 2012, the Free Syrian Army helped him flee the embattled Aleppo and led him across the border to Turkey.[162] In 2013, he was reportedly named head of a new government formed by the Free Syrian Army.[163]

While remaining a celebrity, the cosmonaut shared the fate of many other Syrian refugees settling in Turkey, including restrictions on international mobility. In 2015, he was invited to participate in the XXVIII Planetary Congress of the Association of Space Explorers in Sweden. However, the Swedish consulate general in Istanbul refused to issue him a visa, despite his pledge to stay in Europe for only fifteen days.[164] Despite his proficiency in Russian, seeking refuge there was also out of question. 'My Russian friends often call me and tell me I am most welcome to go there,' Fares explained, 'But I tell them that is impossible as long as their politicians keep propping up the brutal dictator Bashar al-Assad, who is murdering our people.' He claimed that he received invitations from other Western countries, but described them 'as a token to conceal their apathy toward my country and my people'. He was thus partly forced, and partly chose, to remain in Turkey, 'where I can try to at least do something for the liberation of my country'.[165]

Outraged by the violence in Syria, Fares's pacifism waned. In 2012, an interviewer for the Saudi newspaper *Okaz* asked him what he would do if he were charged with bombing al-Assad's palace. Fares responded that he would 'not hesitate for one moment', as this would be a 'heroic and honorable action for all Syrians'. The former cosmonaut added that he had reached the age of 60 and could no longer carry weapons. If he were a young man, he 'would not hesitate to join the rebels and defend my homeland and people'.[166] Fares expressed similar hostility towards his former government to Western audiences. 'I tell Europe if you don't want refugees, then you should help us get rid of this regime,' he told the Associated Press in 2016. 'We in Syria, sacrificed more than a million people and 1.9 million injured,' he added. 'Can we let this be for nothing?'[167]

Despite his restrictions on travel and support of violence, Fares remained a cosmopolitan. In 2015, he gave another interview for the Emirati newspaper *The National*, in which he expressed a similar view as Sultan bin Salman: 'When you go up there, you realise there are no borders, no countries, no nationalities. Just Earth. Mother Earth,' he said. 'We should protect this Earth, as who hurts their mother?' Fares also stressed the transformative experience of leaving Earth: 'The way you think about life in general changes – you start thinking of the bigger picture and you don't let the small things that happen daily bring you down,' he stated. 'There is an inner sense of calmness that stays with you.' Amidst the Syrian Civil War, he was still hopeful, 'Whatever happens, we always have to look forward and rebuild. It is important to remember Syria for its long and rich history, its many pioneers, and how it conquered many frontiers, including the final frontier of space.'[168]

With Syria still ravaged by civil war, however, hopes for new space initiatives shifted again to the Gulf states, which possessed the necessary wealth and stability for ambitious international collaborations in science. This wealth also created jealousies and some of the projects on which it was spent attracted considerable criticism.

# CHAPTER 6

# The Arabs' Final Frontier

In December 2011, the journal *Science* published a news story entitled 'Saudi Universities Offer Cash In Exchange for Academic Prestige.' The author claimed that King Saud University (KSU) in Riyadh and King Abdulaziz University (KAU) in Jeddah were hiring highly-cited foreign scientists as visiting and adjunct professors. In return for tens of thousands of dollars in salaries or grants, these scientists were expected to teach courses, collaborate in research and adopt their Saudi employers as second affiliations. Because of increased visibility, both institutions soared in the international rankings.[1] The news story caused much controversy, with *Science* receiving more than 120 comments and various newspapers around the world reporting on it. Perhaps making a subtle reference to the attacks on 11 September 2001, an article in *The Australian* was entitled 'Saudis recruit to hijack rankings.'[2] The leadership of King Saud University defended its 'international collaborations', promising 'landmark achievements in knowledge production for the benefit of Saudi Arabia and the world'.[3] However, in July 2012, the Saudi king dismissed Abdullah Al-Othman as rector of KSU.[4] In one article, one of Al-Othman's faculty members listed the 'scandal' of the *Science* story, along with 'administrative and financial chaos', among the causes of this dismissal.[5]

The Saudi programs divided the international astronomical community. King Abdulaziz University offered several highly-cited astrophysicists adjunct professorships with an annual salary of $72,000. In return, they were supposed to visit Jeddah for one or two weeks per year, supervise a research group and add KAU as their second affiliation. One of the astrophysicists, George Miley from the Leiden Observatory, initially welcomed the offer as an opportunity to 'build up astronomy' in the kingdom. However, he declined the adjunct professorship after having recognised it as a tool 'to boost the ranking of the university artificially'.[6] Jeremy Mould in Australia was also tempted, but worried whether the joint affiliation would affect the ranking of his main institution, Swinburne University of Technology.[7] In contrast, Ray Carlberg from the University of Toronto considered a similar offer by KAU 'sincere' and accepted it. He submitted a proposal for funding a telescope in the Canadian Arctic and hoped to involve faculty members and students from KAU. Another KAU affiliate, Gerry Gilmore from the University of Cambridge, expressed pragmatism. 'Universities buy people's reputations all the time,' he noted, 'in principle, this is no different from Harvard hiring a prominent researcher.'[8] Numerous other researchers in space science and other fields also affiliated with the university. In 2014, 160 authors on Thomson Reuters's website Highlycited.com were employees of KAU. This was the world's second highest number after the University of California, which represented an entire university system.[9]

Whatever one's views of this exchange of money, expertise and prestige, it demonstrates the extent to which globalisation and capitalism have affected Saudi universities. Whereas KAU's Department of Astronomy mostly recruited and collaborated with Egyptians and Saudis between the 1970s and 1990s, it widened its horizons in the 2000s and 2010s. The university's 'distinguished adjunct professors' in space science included – besides Carlberg and Gilmore – the following: Stanislav Djorgovski from the California Institute of Technology, Jay Frogel from Ohio State University, Luigi Piro from Italy's National Institute for Astrophysics and Luis Rodríguez from the National Autonomous University of Mexico.[10] By the time of the *Science* story, KAU's leadership was also acutely aware of the international rankings and the strategies of other institutions. Adnan Zahed, the university's vice-president for graduate studies and scientific research, stated that his 'program is not different from those in many elite universities around the world from which top scientists continue to receive attractive offers.'[11]

The globalisation of Saudi space science did not mean that national interests had become less important. Instead, national capacity building was part of the investments in global science. Zahed argued that 'KAU is clearly making a sound strategic business investment for the future of our country.'[12] Even Miley, who declined the adjunct professorship, admitted that 'astronomy can be a unique tool for capacity building and global development.'[13] Uwe Becker, a German physicist and member of KSU's Distinguished Scientist Fellowship Program, claimed that his work was an 'effort to strengthen the still relatively weak scientific community in Saudi Arabia, a venture undertaken by many eminent scientists from around the world.' Becker called for an appreciation of these international collaborations 'as they help the Arabic countries to survive'.[14]

While many people argued for Saudi investments in science in general, some emphasised space research in particular. In 2009, Hamad Al-Lohedan, a professor of chemistry at King Saud University, published an article in the *al-Riyāḍ* newspaper entitled 'The importance of establishing the Saudi space agency.' He argued that that 'the space industry today has become extremely important in military, economic and security terms.' He explained that Saudi Arabia needed satellites in order to monitor its borders, waters and airspace. This spacecraft would further benefit the kingdom in terms of communications and in the search for water, oil and other natural resources. Al-Lohedan acknowledged the work of the Arab Satellite Communications Organization, but complained that it had not fulfilled its role in terms of space science and the encouragement of related industries. He said that this organisation, like others, was still relying on others to launch its satellites into orbit. The Saudi professor insisted, however, that satellites were part of 'national security' and 'national sovereignty'. They should not be contracted out to others.[15]

While Al-Lohedan called for a Saudi national rather than a pan-Arab institution, Arabism and Islam were still part of his rhetoric. Perhaps trying to shield himself from religiously-based criticism, he began his article by stating that 'Islam is the religion of openness to the other, of research, investigation, and leadership in everything.'

After quoting several Qur'anic verses, he drew on a regional collective memory. He wrote that 'the early Muslims had carried the banner of research, investigation and exploration in the field of astronomy.' Located 'at the heart of the world' and rich in resources, the Arab region also had the capacity to 'play a vital role in all political, economic and scientific fields', the chemist added. The 'enemies of the nation' had recognised this and had thus pursued a 'policy of divide and rule' that had 'torn the Arab ranks'. Al-Lohedan promised, however, that a Saudi space agency would serve 'Arab security and stability' from the Atlantic Ocean to the Gulf.[16]

Despite his best efforts, Hamad Al-Lohedan struggled to fully convince his audience. In a follow-up article, the chemist acknowledged having received a wide variety of responses. He sensed that there was a 'lack of understanding, ambiguity or limited knowledge of the role and functions of space agencies'. Many of his readers believed that this role was confined to 'sending astronauts and spacecraft to observe the other planets and stars'. The chemist explained again that space agencies 'have other duties and responsibilities related to security, defense, education, research, development, the economy, medicine, climate, pollution and the discovery of natural resources'. Al-Lohedan ended on another nationalist note, writing that the 'establishment of the Saudi space agency would fill a major hole in the skies of the Arab world and increase the kingdom's regional and international weight.'[17]

In Al-Lohedan's proposal, national security was more important than any cosmopolitan aspects of space of space science. While other scientists had crossed national borders themselves or contributed to a borderless view of Earth, he emphasised border control. In his follow-up article, he wrote that 'we can protect our airspace from being violated and our privacy, our borders, our waters and our communications from being penetrated.'[18] The chemist thus seemed convinced by the potential of satellites for total control. Arguably, this belief in the power of having a view from above had a long history. The British Royal Air Force and subsequently the United States Air Force were convinced that planes and drones were among the best tools for policing large areas of the Middle East.[19] Al-Lohedan's articles suggested that at least some Saudis had adopted a similar conviction regarding space technology by the early 21st century.

Al-Lohedan's proposal remained just that. However, other scientists continued to promote ambitious projects in space science, even as the Arab region faced new conflicts and drops in oil prices after 2011. In March of that year – just a few months into the Arab Spring – Hisham Abad from Alfaisal University presented the idea of sending a Saudi citizen to Mars at TEDxNajd, a conference in Riyadh.[20] In 2014, he and two co-authors from King Saud and Dar Al Uloom universities in Riyadh described the project in *al-Maʿrifah*, meaning 'Knowledge', a magazine published by the Saudi Ministry of Education. Abad and his peers argued that Saudi Arabia was facing 'great challenges', including the diversification of its economy and an 'education crisis'. In order to 'raise living standards' and 'compete in the global economy', the authors recommended a Saudi Mars mission using the Apollo program as a model. They argued that this mission would contribute to progress in various fields of science and technology in the kingdom, as Apollo had done in America during the 1970s.[21]

The response to this proposal of a Saudi Mars mission was limited, and Hisham Abad soon lost his position at Alfaisal University. In March 2014, the institution was still congratulating him on the publication of his article.[22] However, by that time he had fallen out with his supervisors. They included his dean and fellow physicist Nouredine Zettili and the provost Faisal Al Mubarak, whom Abad accused of 'corruption' and 'nepotism' (*wāsiṭah*) in hiring practices.[23] The university terminated his employment, arguing that he had failed to 'abide by applicable local laws, rules and regulations and custom'. The university also told him that he had failed to 'carry out generally accepted academic professional standards and practices as a faculty member'.[24]

Although the Saudi space agency and Mars mission remained unrealised, Arab governments had increasingly come to see space science and technology as important parts of their national futures by the early 21st century. New communications satellites belonging to organisations such as Arabsat and Nilesat were launched as well as various research satellites. The latter included the *Saudisat* series developed by the King Abdulaziz City for Science and Technology, *Maroc-Tubsat,* Algeria's *Alsat-1* and the *EgyptSat* series. While communications satellites were often commercial and used for television channels, research satellites served Arab states in various other ways. These included remote sensing for development planning and disaster management. In addition, research satellites often had educational purposes, as participation in their development provided nationals with training in advanced technology.[25]

Although they were civilian rather than military, at least some of these satellites were also intended to contribute to national security. *EgyptSat-1,* launched in 2007, for the first time provided one of Israel's neighbours with 'comparable intelligence-gathering capabilities'.[26] Medhat Mokhtar, head of the National Authority for Remote Sensing and Space Sciences (NARSS), commented on the launch of *EgyptSat-2* in 2014: 'The new satellite will primarily help advance three main issues: namely, the conclusion of studies to develop the Suez Canal axis, the development of projects for the southeastern region of Egypt and the monitoring of the Renaissance Dam's effects on the water security of Egypt.' Egyptian officials were concerned that Ethiopia's Renaissance Dam was impacting the Nile Valley by trapping water and altering its flow rate. Mokhtar also added that the 'satellite will not serve any military purposes, but its uses do fall within the purview of national security through the monitoring of Egypt's borders with its neighbors and the detection of illegal activities along our borders.'[27]

In contrast to Arabsat, the satellites launched in the 21st century mostly belonged to individual states rather than to the Arab League as a whole. Nevertheless, several of these projects still drew on a common Arab heritage. *Maroc-Tubsat,* which was launched in 2001, was nicknamed *Zarkaa Al Yamama.* It thus referred to a legendary, Cassandra-like figure from pre-Islamic Arabia, a blue-eyed woman with keen sight and an ability to predict events.[28] In 2010, Qatar, which had used international satellites to broadcast Al Jazeera since the 1990s, established its own company, Es'hailSat. It was named after Canopus (*suhayl*), an important star in Bedouin lore. Its annual appearance in October announces 'clement weather and a joyful season in the Middle East'. On Qatar's National Day in 2013, *Es'hail-1* entered service by transmitting beIN Sports and Al Jazeera.[29]

Beyond the human, financial and symbolic resources of Arab countries, satellite projects also relied on global connections. These connections transcended the relationships with former colonial powers and Cold War allies. In 2000, *Saudisat-1A* and *-1B* were launched not on NASA's space shuttle, nor on a European Ariane rocket, but on a Russian rocket from Baikonur in Kazakhstan.[30] The 'Tub' in *Maroc-Tubsat* stands for Technische Universität Berlin rather than for a French institution. This Moroccan–German satellite was also launched on a Russian Zenit rocket from Baikonur in Kazakhstan rather than on an Ariane rocket from French Guiana. Algeria's *Alsat-1* was designed by Surrey Satellite Technology in Britain, and Qatar's *Es'hail-1* was jointly owned and operated by France-based Eutelsat.[31]

Although Egypt's remote sensing satellites were supposed to contribute to national security, their construction and launch depended on cooperation with foreign partners. The National Authority for Remote Sensing and Space Sciences built *EgyptSat-1* together with the Yuzhnoye Design Bureau in Ukraine; it was launched on a Dnepr rocket from Baikonur in Kazakhstan.[32] The failure of the satellite in 2010 became the 'joke of the day' in Egypt, as one blogger put it,[33] adding to popular criticism of the government. *EgyptSat-2* was built by the Russian Rocket and Space Corporation Energia and launched from a Soyuz-U rocket, also from Baikonur. However, in 2015, less than a year after its launch, Egypt also lost control of this second satellite.[34] To compensate for this loss in the short term, Egypt reportedly considered buying Russian satellite images.[35] However, in 2017, the Egyptian government also sought to cooperate with China in order to continue its satellite program.[36]

Other Arab countries also continued to rely on foreign partners for technology related to national security, especially missiles. In 1988, Saudi Arabia announced that it had received Chinese Dongfeng missiles, meaning 'East Wind', with a range of 3,000 kilometres.[37] The Gulf Cooperation Council (GCC) countries also purchased Patriot and Terminal High-Altitude Area Defense (THAAD) systems from the United States at various points during the late 20th and early 21st centuries. These systems were partly aimed at protecting the Arab monarchies against Iranian missiles.[38] Following its war with Iraq during the 1980s, Iran developed a series of missiles named

Shahab, meaning 'Meteor'. With Russian help, the Islamic Republic established a space agency in 2004. In 2005 and 2009, Iran launched its first two satellites, named *Sina* and *Omid*, meaning 'Hope'. One Iranian official described his country's space program as 'riding high for peace and pride'.[39] Nevertheless, the Iranian launchers, together with the country's nuclear program, were still causing considerable concern in other Gulf countries as well as Israel and America.

Perhaps partly in response to the Iranian space endeavours, the United Arab Emirates developed a similarly ambitious program. This program was also connected to nationalism and cosmopolitanism, and relied on massive state support and international collaborations. The UAE was arguably one of the most successful Arab unity projects in modern times. It lasted much longer than the United Arab Republic of Egypt and Syria and the Arab Federation of Iraq and Jordan, which were formed during the late 1950s. In addition to Arabism, diverse non-Arab communities and openness towards the world characterised Emirati cities, especially 'global Dubai'.[40] Many communities are, however, gated and segregated.[41] In parallel, 'nationalisation' (*tawṭīn*) programs seek to promote Emiratis and preserve their privileges.[42] Nationalism, regionalism and cosmopolitanism co-exist, albeit in a state of tension with one another, and supported visions of an Emirati future in space.

## HOPE

By the 2000s, the United Arab Emirates, and Dubai in particular, had associated themselves with a distinct vision of the future. The artists Sophia Al-Maria and Fatima Al Qadiri coined the term 'Gulf futurism' in order to describe and critique the combination of capitalism and religious conservatism that shaped the UAE and its neighbours.[43] Whether dystopian or utopian, the deserts and hypermodern architecture attracted makers of science fiction films, including space operas. Michael Winterbottom shot his 2003 film *Code 46* partly in Dubai.[44] Several episodes of *Star Wars* were filmed in Tunisia, where the town Tataouine provided the name for the desert planet Tatooine. However, after 2011, the area around Tataouine featured in actual fights against jihadists (rather than Jedi).[45]

Therefore, parts of *The Force Awakens* were shot in Abu Dhabi instead. The government of this emirate lured film makers not just with its greater political stability, but also with financial incentives. These incentives were perhaps part of a bigger strategy to diversify the oil-driven economy through movie productions and tourism.[46]

In addition to attracting international productions, the United Arab Emirates inspired local science fiction. In 2012, Noura Al Noman, a former teacher and government censor from Sharjah, published *Ajwan*, an Arabic novel about the abduction of a child by aliens.[47] The author said, 'diversity in the UAE is almost like a science fiction story, where the world has become one large nation with so many ethnicities and faiths all interacting and living side by side.'[48] The following year, Dubai-born Pakistani director S. A. Zaidi and Emirati producer Ghanem Ghubash released *The Sons of Two Suns*, an apocalyptic movie in which two suns slowly burn Earth. In 2016, the duo released *Aerials*, in which alien spaceships arrive and shut down Dubai.[49]

Although these science fiction productions are relatively recent, enthusiasm for global space exploration in the Emirates began even before they formed a federal state in 1971. As early as 1964, Dubai issued four stamps 'honouring astronauts' and celebrating 'outer space achievements', specifically the Ranger 7 mission. In 1966, Umm al-Quwain issued five stamps showing different satellites, including *Telstar-1* and *TIROS-1*, in Earth orbit. Subsequent stamps from Dubai featured the 'US Gemini space project' and its 'rendezvous in space'. Other issues showed the first Moon landing and the 'Outer Space Telecommunications Congress' in Paris in 1971. Although many of these stamps, known as dune issues, were primarily sold to foreign collectors, some of them ended up on display at local museums. Furthermore, in 1973, visiting Apollo astronauts presented a lunar rock as a gift to the first president of the UAE, Sheikh Zayed bin Sultan, who decided to place it in a museum in Al Ain. There, the rock became part of a section on space exploration that was juxtaposed with local artefacts and photographs from the middle of the 20th century. This juxtaposition thus created a contrast between a 'simple, pre-modern past and a hypermodern present and future'.[50]

Having participated in Arabsat since the 1970s, the United Arab Emirates began investing in a second regional satellite company in the late 1990s. In 1997, the Emirates Telecommunications Corporation (Etisalat), in partnership with Arabsat and Qatar Telecom, established the mobile satellite phone provider Thuraya with headquarters in Dubai. This company was named after the Arabic designation for the star cluster Pleiades. The annual ascendance of this star over Arabia occurs in late October, two weeks after that of Canopus, ushering in the rain season.[51] At a cost of almost $1 billion, Boeing developed satellites for Thuraya that were launched in 2000 and 2003.[52] Encouraged by the success of Thuraya, Abu Dhabi established Al Yah Satellite Communications Company (Yahsat) in 2007. Its first satellite, constructed by EADS Astrium and Thales Alenia Space, was launched on an Ariane rocket in 2011.[53]

Partly to save fuel, the first Thuraya and Yahsat spacecraft were launched close to the equator, from the Pacific Ocean and French Guiana, respectively. However, the UAE also sought to develop spaceports on its territory, complementing its growing hubs for sea

**Figure 6.1** Stamps on display in the Sheikh Saeed Al Maktoum House, Dubai, 2016 (courtesy of the author)

and air transport. In 2006, Space Adventures announced plans to build a $265 million spaceport in Ras al-Khaimah. While these plans made little progress, Abu Dhabi's Aabar Investments acquired a 32 per cent stake in Virgin Galactic together with rights to launch tourism and research flights for $280 million.[54] In 2007, Ibrahim Sharaf, an Emirati entrepreneur, paid $200,000 for a flight with Virgin Galactic that was scheduled for two years later.[55] However, the company's first commercial space flight was delayed on several occasions, including the failure of a test flight in 2014. In the meantime, space tourism projects proceeded on the ground, including an $8 million observatory on a mountain in Ras al-Khaimah. With guest rooms for up to 70 people, the observatory was intended to be an 'astronomical resort', said Hasan Hariri, head of the Dubai Astronomy Group in 2014. 'We don't want to make people travel so far to go to an observatory,' he added. 'This is part of the way we can bring astronomy as close to the people as possible.'[56]

Emirati projects went ahead despite some religious objections to the risks of human space flights. In 2014, the UAE's General Authority of Islamic Affairs and Endowments issued a fatwa prohibiting a one-way journey to Mars, if the chances of death were greater than survival. Participation on this journey would be punished the same as suicide in the Hereafter. This fatwa came after Mars One, an organisation based in the Netherlands, had proposed no-return flights to Mars in order to establish a permanent human colony.[57] Perhaps as a result of this prohibition, Arab participation in Mars One was limited although Mohammed Sallam, the only Egyptian Mars One candidate, expressed adherence to Islamic beliefs. He told Al Arabiya in 2015 that he hoped to become the first person to 'carry the Qur'an to another planet', to fast and to pray in the direction of Earth. Sallam, who worked for an insurance company, also saw himself as a representative of humanity and an Egyptian patriot. In addition to the Qur'an, he wished to take his country's flag and Shadia's song 'Oh my beloved, oh Egypt' (*Yā ḥabibtī yā Maṣr*) to Mars. Finally, Sallam hoped to grow beans, falafel and *mulūkhīyah* – a vegetable and his 'favourite dish' – there.[58]

With increasing investments in the space industry, the UAE offered opportunities for scientists from other Arab countries.

In 1998, the Emirates Center for Strategic Studies and Research in Abu Dhabi published a book by Farouk El-Baz entitled *The Arab World and Space Research: Where Do We Stand?*[59] The Algerian Nidhal Guessoum took up a position as assistant professor at the American University of Sharjah in 2000 and was promoted to professor in 2008. The same university also hired Hazim El-Baz, one of the Farouk's brothers, as a faculty member in engineering. Hazim had earned a degree in aeronautical engineering at Cairo University before pursuing postgraduate studies in Missouri like his more famous sibling. The Iraqi Hamid Al-Naimiy left Jordan's Al al-Bayt University in 2001 to become head of the physics department at the UAE University in Al Ain. From there, he moved to the University of Sharjah, where he rose from dean to chancellor. In 2008, he made Sharjah's ruler Sultan al-Qasimi honorary president of the Arab Union for Astronomy and Space Sciences (AUASS).[60]

While Arabsat, Thuraya and the AUASS were regional projects, the UAE also invested in more explicitly national endeavours. In 2006, Mohammed bin Rashid, the ruler of Dubai, decreed the establishment of the Emirates Institution for Advanced Science and Technology (EIAST). This institution subsequently ordered the manufacture of two Earth observation satellites, *DubaiSat-1* and *-2*, in South Korea. In 2009 and 2013, respectively, Dnepr rockets took them from Kazakhstan and Russia into orbit. In order to participate in the production of both satellites, the UAE sent scientists to South Korea. However, the EIAST subsequently announced that a third satellite would be constructed entirely by Emirati engineers.[61] It was first named *DubaiSat-3*, but then renamed *KhalifaSat* after the ruler of Abu Dhabi. This name change, like the one from Burj Dubai to Burj Khalifa, occurred after Abu Dhabi had bailed out its neighbour financially in 2009. In 2014, the UAE Space Agency was founded in Abu Dhabi in addition to the EIAST in Dubai. By 2015, the development of a space industry had become 'a primary national objective' in the UAE.[62]

In pursuit of funding during a period of lower oil prices, the Emirates Institution for Advanced Science and Technology described space science as crucial for national development. Renamed the Mohammed bin Rashid Space Centre (MBRSC) in 2015, the institution sought to align itself with the UAE Vision 2021 for 'a competitive knowledge-based economy that is fuelled by creativity and innovation in the science and technology sector'. In support of this vision, the MBRSC, together with engineering students at the American University of Sharjah, developed *Nayif-1*, the UAE's first nanosatellite. Meaning 'one that soars high above the ground', this satellite was supposed to symbolise the 'importance of education in the ascent of a nation'.[63]

In parallel to *Nayif-1*, Emirati scientists used their experience of collaborating with their Korean colleagues to embark on an even more ambitious project: the Emirates Mars Mission. In collaboration with the University of Colorado Boulder, long a 'world center for space and atmospheric science',[64] the MBRSC and the UAE Space Agency developed an orbiter to study the Martian atmosphere. The planned orbiter was unmanned, which was most likely due to the complexity of manned spaceflight rather than the earlier Emirati fatwa. Coordinating with NASA's Mars Exploration Program Analysis

**Figure 6.2** Mohammed bin Rashid Space Centre, Dubai, 2016 (courtesy of the author)

Group, Emirati scientists sought to 'complement the current international efforts to study the Red Planet and also provide unique insights to fill the gaps in human knowledge'.[65]

Making the Emirates Mars Mission a global project, the UAE benefitted not just from American and Korean expertise. The country also agreed with Mitsubishi Heavy Industries to launch their probe from Japan. At the same time, the orbiter's instruments were supposed to take the Emirati national brand to Mars. An Emirates eXploration Imager (EXI) sought to make visible the properties of ice and dust aerosols and the abundance of ozone in Mars's atmosphere. The Emirates Mars Ultraviolet Spectrometer was intended to measure the global characteristics of hydrogen and oxygen coronae. Finally, the Emirates Mars InfraRed Spectrometer (EMIRS) would measure the global thermal structure and the abundance of ice and vapour.[66]

As the widespread usage of the name 'Emirates' suggests, Emirati and Arab nationalism was a strong component of this project. In addition, the Emirates Mars Mission, like *KhalifaSat*, was linked to pride in, and the prestige of, the ruling families. The probe was named *Al Amal*, meaning 'hope'. It was scheduled for launch in 2020, the year of the Expo in Dubai. The spacecraft was supposed to reach Mars in 2021, coinciding with the fiftieth anniversary of the United Arab Emirates. In 2015, Mohammed bin Rashid stated that one of the mission's messages was 'to our Arab brethren: that nothing is impossible, and that we can compete with the greatest of nations in the race for knowledge.'[67] 'Sheikh Zayed was the hope of the UAE and the UAE is the hope of the region,' said Sheikh Mohammed about the first president of the Emirates. He added, 'Our generation is the hope of Arabs and Muslims, so the choice of the name for the probe is Hope.'[68] The official announcement made no reference to the Iranian satellite *Omid*, meaning 'Hope', or to a science fiction story by Raouf Wasfi that also featured a spacecraft named *al-Amal*.[69]

Like previous Arab space ventures, the Emirates Mars Mission combined nationalism and cosmopolitanism. Sheikh Mohammed stated that the mission's 'first message is for the world: that Arab civilisation once played a great role in contributing to human knowledge, and will play that role again.'[70] Omran Sharaf, one of the project managers, said that the mission 'is not about reaching Mars but about inspiring a whole new generation and transforming the way youth think within the region.' He added, 'The goal here is hope, for humanity, for the region, for youth in countries with lots of conflict.'[71] The Emirati researchers planned to make all their data freely available to 200 universities and research institutes. These data would 'give the international science community a deeper and richer understanding of the Martian atmosphere', said one Sharaf's deputies.[72] Sarah Amiri, another Emirati manager, expressed a view of Earth and humanity similar to those of Sultan bin Salman and Muhammed Faris. 'Space is very humbling, and an image of Earth without any boundaries, the actual conflicts and how small they appear,' she said, 'it makes you think more about putting our differences aside and treating each other as humans and one species.'[73]

The Emirates Mars Mission did not remain at the end of the country's ambitions to explore the red planet. In February 2017, Mohammed bin Rashid and Mohammed bin Zayed Al Nahyan, the crown prince of Abu Dhabi, announced the Mars 2117 project. It aimed to build a city on Mars within a hundred years. Cosmopolitanism and nationalism were also fully part of this project. On one hand, Mohammed bin Rashid expressed a global consciousness. 'The landing of people on other planets has been a longtime dream for humans,' said the ruler of Dubai. 'Our aim is that the UAE will spearhead international efforts to make this dream a reality.' On the other hand, Mohammed bin Rashid again emphasised national capacity building. 'The 2117 Mars initiative is a long-term project, which will first help develop our education, universities and research centres that will empower young Emiratis to enter all disciplines of scientific research fields.'[74]

In addition to pursuing global and national aims, the Mars 2117 project sought to contribute to national and regional security. Preparing for life on the red planet involved overcoming challenges of water, food and energy security, which the UAE also expected on Earth in the future.[75] Other terrestrial challenges addressed by Mars 2117 were more demographic and ideological. 'In the UAE, we live in a rough neighborhood,' said Saeed Al Gergawi, the director of the project at the Mohammed bin Rashid Space Centre in May 2017.

'Our neighborhood has over 100 million youth, with over 35 per cent unemployment,' he told an audience at the Humans To Mars Summit in Washington, DC. This high rate was one cause of radicalisation and terrorism, Al Gergawi explained. One of the aims of the Mars 2117 project was to direct the energy of Middle Eastern young people away from violence and towards education, science and technology.[76]

Within a few years, the space ambitions of the UAE government had thus far exceeded those of its Egyptian and Saudi counterparts, despite their longer track records. After the loss of *EgyptSat-1* and *-2*, the Egyptian space program had become a laughing stock,[77] irrespective of celebrity scientists such as Farouk El-Baz. On 1 April 2017, the website *Egyptian Streets* published an article entitled 'Egypt Plans to Build First City on Mars.' The website stated, 'Egypt's government has announced plans to build a city on Mars by 2050, heralding a new era of space exploration by the Egyptians.' This city would be a 'new home for mankind, one that is hopefully free of politics and instead focused on our future'. The website later explained that the story had been published to 'mark April Fool's Day'. The website felt this explanation was necessary 'after a leading private Egyptian newspaper initially reported this as true. However, we do one day hope Egypt will expand its focus on science and research, and maybe even launch a space agency as has been planned for several years.'[78]

The UAE's Mars 2117 project was not a joke. However, the recent Emirati space initiatives were unexpected even for some scientists in the country. After promoting space exploration for years, Nidhal Guessoum described the announcement of the Emirates Mars Mission and the creation of the UAE Space Agency as a 'pleasant surprise'. He said in 2014 that Arab government officials and members of the public saw space exploration as 'a luxury, if not a useless endeavor, that certainly does not deserve and cannot justify billions of dollars in spending'. He praised the UAE president for 'aiming, through space exploration, at developing national expertise and capabilities'. Nevertheless, he criticised recent endeavours for the 'near-total inexistence of coordination or even collaboration' between Gulf states or at the 'pan-Arab level'. The Algerian thus repeated calls for a pan-Arab space agency that would reduce the 'costs of launching and operating satellites and allow many countries to benefit from the images and data collected by the various satellites, instead of needlessly duplicating efforts and multiplying spending'.[79]

Although a pan-Arab space agency remained unrealised, Arabism was not absent from the Emiratis' ambitions. While the Emirates Mars Mission had mainly East Asian and American partners, the Emirati scientists were also engaged in some regional collaborations. In April 2017, the UAE Space Agency signed a memorandum of cooperation with its Algerian counterpart, which covered an exchange of expertise. On that occasion, Mohammed Al Ahbabi, director general of the UAE Space Agency stressed 'the necessity to unify Arab efforts to revive the glories of the Arab ancestors who were among the first innovators in the space and astronomy fields'.[80]

## GLOBAL NETWORKS

Whether the Algerian–Emirati memorandum of cooperation will develop into a wider regional partnership remains to be seen. That Arab states at times cultivated their own international connections rather than channelling resources through a pan-Arab agency may, however, also have some advantages. A lack of regional cooperation might make projects more expensive, but it will also potentially reduce the impact of failures on the region as a whole. A pertinent example of an unrealised regional project is the Network of Oriental Robotic Telescopes, which French and Moroccan astronomers, including Samir Kadiri and Zouhair Benkhaldoun, developed during the 1990s. François and Monique Querci, who were based in Toulouse, had become interested in automated telescopes during the 1980s and were looking for better observation locations. In 1991, they presented an 'automated station project' with three one-metre telescopes for observing variable stars together with comparison stars. The Quercis suggested the Oukaïmeden peak in Morocco as a location. Around 3,000 metres high, it offered more nights with clear skies than many mountains of a similar altitude in Europe. Claiming the support of around fifty astronomers, the Quercis applied to the French Institut National des Sciences de l'Univers and the Moroccan Centre National pour la Recherche Scientifique et Technique for funding.[81]

In the next few years, François and Monique Querci extended their project eastwards. Using data links with satellites, they selected automated telescopes that would require 'no permanent human presence'. Advertising their technology as 'simple and robust',[82] they promised it to be replicable in different countries. With meteorological data from geostationary satellites, they prospected dry mountains in southern Algeria, the Sinai Peninsula and southwest Arabia. In 1994, the Quercis, Samir Kadiri and Zouhair Benkhaldoun presented the Network of Oriental Robotic Telescopes at a meeting of the Astronomical Society of the Pacific. Their stated objective were 'non-stop observations' by collaborating with the Global Network of Automatic Telescopes (GNAT).[83] The NORT would observe variable stars with telescopes based between China and Morocco. With the sun rising in the east, the GNAT would continue the observations with telescopes based further to the west.

With the support of the Institut National des Sciences de l'Univers and the Alpes-de-Haute-Provence department, the Quercis planned the central node of the NORT to be in France. Engineers at the Haute-Provence Observatory would design a typical station, including a weather unit, telescopes, photometers and antennas for satellite communications. Ten or twelve of these observation stations would be required between China and Morocco to follow a star without interruption. The Haute-Provence Observatory would offer training in this technology and provide universities in Arab and Muslim countries with small telescopes for teaching.[84] In search of donors and collaborators, the Quercis promoted the NORT through the International Astronomical Union and the United Nations. In 1994, François presented the project at a meeting of the Working Group for the Worldwide Development of Astronomy as part of the IAU's General Assembly.[85] In the next few years, the Quercis proposed the project again at different UN/European Space Agency Workshops on Basic Space Science. In 1996, the UN Committee on the Peaceful Uses of Outer Space selected the NORT as a follow-up project for the workshop series.

With backing from France and the United Nations, the NORT sparked a lot of interest among astronomers across the Middle East and even in South Asia. The Arab Union for Astronomy and Space

Sciences led by Khalil Konsul and Hamid Al-Naimiy adopted the NORT at a conference in Amman in 1998. The two astronomers, together with the Quercis, subsequently proposed to 'establish, as a subgroup, an Arab Project for NORT'. They argued that it would 'contribute significantly to capacity building in basic space sciences in the Arab World'.[86] In 2000, Ahmad Al-Mousli from the Syrian General Organization of Remote Sensing gave a presentation on 'Astronomical Site Testing with a Meteorological Satellite' at the Ninth UN/ESA Workshop on Basic Space Science in Toulouse. Al-Mousli described this testing as 'the first step for the implementation of a network of robotic telescopes'.[87] At the same workshop, scientists from Karachi presented 'the case for Pakistan to join NORT'.[88]

Despite the enthusiasm of Middle Eastern and South Asian astronomers, the NORT faced great challenges. The Arab Union for Astronomy and Space Sciences, which suffered from a shortage of funding, could barely contribute financially. Other difficulties included political instability and a lack of infrastructure in some of the locations the Quercis had identified. In the 1990s, François Querci arrived in Yemen, whose mountains he considered 'the most suitable place in the Arabian Peninsula for the installation of a Regional Observatory'. However, he was 'astonished, finding only some old people concerned with a little mathematical astronomy on the beginning of Ramadan or about agricultural seasons, but finding no teaching of astrophysics either by universities or by schools'! Querci also diagnosed that there was 'not a single telescope working in any scientific or educational establishment in Yemen'. In 2001, he was still hoping for the opening of a small observatory with a twenty-centimetre telescope financed by the French Embassy in Sana'a.[89]

In some wealthier Arab countries François Querci also found suitable observatories and astronomers to be scarce. In 2002, he gave a lecture on his project at Algeria's Centre de Recherche en Astronomie, Astrophysique et Géophysique (CRAAG). Another lecture at a university also provided an opportunity to 'discuss various problems of teaching of astronomy and astrophysics and of the Algerian participation in the NORT.'[90] After the exodus of scientists such as Nouredine Zettili and Nidhal Guessoum during the

Algerian civil war, almost an entire generation of leading astrophysicists was missing. One of those who remained, Jamal Mimouni of Contantine, complained that CRAAG was mostly a 'geophysics center' with 'weak output', 'no peer review' and without a 'direct link with the academic world'.[91] Mimouni, who had received his doctorate from the University of Pennsylvania in 1985, was hopeful at first. In a 2008 article, he stated that a new astrophysics doctoral school together with a proposed observatory in the Aurès region should 'put Algeria on the global astronomical map, where it is singularly absent'.[92] In 2010, 'administrative hurdles' led to the freezing of the École Doctorale d'Astrophysique.[93] By that time, François Querci had retired, effectively ending his quest to establish the Network of Oriental Robotic Telescopes. One of his final efforts was to search for a telescope site in Libya during the mid-2000s.[94]

In contrast to the failure of the Quercis' regional project, a number of global networks that received sufficient support from a single and more stable Arab country succeeded. A pertinent example is the collaboration on satellite laser ranging between scientists in Helwan and Prague. These scientists contributed to geophysics research, including plate tectonics. Their collaboration began as part of the Interkosmos Laser Ranging Network, which included stations in East Germany and Poland,[95] but which survived the end of Cold War. As one of very few such stations in Africa for decades, the Helwan Satellite Laser Ranging Station remained an attractive partner for the Czech Technical University in Prague, which first set it up in 1974. In 1992, the station became part of a European Laser Consortium and, in 2000, of the International Laser Ranging Services.[96] In the mid-2000s, the Czechs were paying around $34,000 and the Egyptians $25,000 annually to operate the station.[97]

The reluctance to pool resources in some recent space science projects might also result in a greater diversity of ventures, thus spreading the risks of failure. The Arab region does not need to put all of its 'hope' for advances in space research in the Emirates Mars Mission. Independently from his Emirati peers' international collaborations, the Qatari astronomer Khalid Alsubai has participated in other global networks. In 2010, he was part of the Microlensing Network for the Detection of Small Terrestrial Exoplanets

(MiNDSTEp). Connecting telescopes around the world, this network in turn collaborated with others such as the Microlensing Follow-Up Network (μFUN), the Probing Lensing Anomalies Network (PLANET) and RoboNet.[98] Thanks to his networking and funding from Qatar Foundation, Alsubai was able in 2016 to discover the three new exoplanets that he had already expected three years earlier: Qatar-3b, -4b and -5b.[99]

The Oukaïmeden Observatory in Morocco did not wait for the success of a single cooperative venture either, such as the Network of Oriental Robotic Telescopes. In the 1990s, Zouhair Benkhaldoun and Samir Kadiri participated in the Global Oscillation Network Group (GONG) project, which sought to study the sun continuously.[100] In the 2010s, Benkhaldoun and his student Khalid Barkaoui also collaborated with Michaël Gillon from the University of Liège in Belgium. In 2016, they installed at Oukaïmeden the TRAPPIST-North, which stood for Transiting Planets and Planetesimals Small Telescope. This instrument was connected to TRAPPIST-South in Chile. Together with international colleagues, the Moroccan scientists contributed to the discovery of planets orbiting the ultracool dwarf star TRAPPIST-1. This was followed by additional observations by the Spitzer Space Telescope and other instruments. On that basis, Gillon, Barkaoui, Benkhaldoun and others published an article in the journal *Nature* about the discovery of 'seven temperate terrestrial planets' around TRAPPIST-1. The astronomers found that the planets 'have equilibrium temperatures low enough to make possible the presence of liquid water on their surfaces'.[101]

Like other major successes in Arab space science, the discovery of planets around TRAPPIST-1 had national, regional and global dimensions. In Morocco, 'the discovery opens the door' for finding more planets, said Barkaoui, who was involved in both observation and data reduction.[102] Regionally, the thirty co-authors of the *Nature* article included Yaseen Almleaky of King Abdulaziz University and the King Abdullah Centre for Crescent Observations and Astronomy in Saudi Arabia. Globally, the other co-authors were affiliated with institutions in Britain, France, South Africa, Switzerland and the United States.[103] The discovery of seven Earth-like planets orbiting the same star was so spectacular that it was covered by major news media in many languages. Barkaoui claimed, 'Oukaimeden is now known throughout the world.'[104]

While the discovery was a source of pride, Barkaoui's teacher Zouhair Benkhaldoun also found it humbling. Perhaps even more than the view of Earth from space, the view of other Earths in space had cosmopolitan potential. It allowed astronomers to transcend not just nations and continents, but even the planet as whole. Benkhaldoun said that his discovery 'allows us to stop thinking that we are the centre of the universe: if your form of life exists someplace else, it means that there is not one single point of creation, which is very important from a philosophical point of view.'[105] To be sure, this view is far from 'citizenship' of the entire cosmos, but it may be a step towards it.

## EPILOGUE: SPACE HERITAGE AND THE ARAB WORLD

If the idea of a medieval Golden Age of Arabic astronomy encouraged quests for a scientific renaissance, 19th- and 20th-century Arab space projects could provide a similar legacy. On gaining the status of national, regional and global heritage, their memory could become a symbolic resource for future space projects. French museums such as the Musée de l'Air et de l'Espace to the north of Paris preserved and displayed artefacts from its rocket program in Algeria between the 1940s and 1960s. Similarly, the Apollo program provided many objects for a growing industry concerned with space heritage, including touring exhibitions. Farouk El-Baz contributed to this industry through his work for both NASA and the National Air and Space Museum in Washington. Many other Arabs arguably derived scientific inspiration even just by visiting these institutions.

Some sites within the Arab region are, of course, still at the frontiers of space heritage. However, they have still attracted the attention of tourists and foreign researchers. The *Star Wars* sets in Tunisia were a popular attraction until political instability after 2011 reduced the number of visitors.[106] I visited the sets on an organised tour for Arabic students at the Institut Bourguiba des Langues Vivantes in 2006. Less known at the time were the remains of the French rocket program near Hammaguir in Algeria. However, in the

2000s, researchers from France[107] and even as far away as Australia began studying them as part of global space heritage. Alice Gorman published an article on 'the archaeology of space exploration' in 2009. She studied the interaction between French rocket scientists and engineers and Tuareg people in order to highlight the 'contributions of "Third" and "Fourth" world people' to space exploration during the Cold War.[108]

While film sets and rockets were perhaps more attractive to many tourists, observatory buildings also attracted more attention from astronomy historians and heritage professionals at the beginning of the 21st century. Around 2001, French and Algerian researchers began studying the Algiers Observatory as part of an 'inventory' of French astronomical heritage along with the observatories in Bordeaux, Lyon, Marseille, Meudon and Strasbourg, among others.[109] Although this inventory was primarily a French project, it gained significant support from the Algerian Centre de Recherche en Astronomie, Astrophysique et Géophysique. In a 2015 article, Françoise Le Guet Tully from the Côte d'Azur Observatory and Hamid Sadsaoud from CRAAG compared the Algiers and Nice Observatories as two 'astronomical heritage sites' on the Mediterranean coast.[110]

For some heritage initiatives, just like space science projects, national interests and cosmopolitanism went together. In 2010, the Egyptian government applied to UNESCO for the Helwan Observatory to be listed as a world heritage site. Arguing for the observatory's 'outstanding universal value', the Egyptians wrote that the Reynolds reflector 'was the first telescope in the world to record Halley's Comet in 1909 and 1911'. In addition, the telescope 'helped the astronomical community in the development of the modern galaxy classification'. The application also stated that the observatory was 'unique among other observatories in the world' in calculating prayer times, the direction of Mecca and the beginning of lunar months. Finally, the Egyptian government compared the Helwan Observatory to Maritime Greenwich and the Beijing Ancient Observatory, two existing world heritage sites.[111]

Nevertheless, few modern astronomical buildings and instruments were afforded the same heritage status as medieval astrolabes, which were considered to be icons of Islamic civilisation and preserved in

**Figure 6.3** The author on the Star Wars set in Tunisia in 2006 (courtesy of the author)

institutions such as Doha's Museum of Islamic Art. However, modern observatories increasingly became the object of conservation efforts. In 2012, a student from a media class at the American University of Beirut produced a short documentary film entitled *The Lee Observatory: A Staple of AUB Left to Rust.*[112] In 2016, when the university celebrated its 150th anniversary, Sevana Sarraf, an engineering student, circulated a petition to turn the Lee Observatory into a space museum. In under two weeks she had collected more than 400 signatures. Fadlo Khuri, AUB's president, doubted whether the number of visitors would be sufficient to justify the expense. Sarraf claimed that her efforts not just revealed how much 'people do want' this museum, 'but also how abandoned of a place the observatory is, and how it really badly needs to be recovered'.[113]

Although most modern astronomical heritage projects sought to preserve long-lasting and well-known institutions, some also focused on short-lived, and largely forgotten, episodes. A pertinent example is the 2012 documentary film *The Lebanese Rocket Society*, which commemorated the rocket program at Haigazian College and the Lebanese military during the 1960s. Memories of this program had faded during the Civil War, but Lebanese filmmakers Joana Hadjithomas and Khalil Joreige recovered enough photographs and other materials to present the program to international audiences. For the fiftieth anniversary of the first rocket launch, the filmmakers produced an art installation for the Sharjah Biennial 10 in 2011. The main focus of the installation was a replica of a seven-metre-long Cedar 3 rocket. From inside the wooden sculpture, a loudspeaker played sounds from the 1960s. This 'Golden Record of the Lebanese Rocket Society' was itself inspired by the Voyager Golden Record. These were discs with sounds and images from Earth that spacecraft were carrying beyond the solar system as messages from humankind.[114] With the rocket situated in front of the Sharjah Art Museum, the replica may have lent some inspiration to the Emirates Mars Mission.

Not only buildings and artefacts, but also the scientists themselves were part of this modern astronomical heritage. In 2002, the Centre de Recherche en Astronomie, Astrophysique et Géophysique started a series of articles on 'figures of the past' in its newsletter. This series paid 'tribute to distinguished scientists that marked the life of our institution, which is now more than a century old'. The first tribute was to Jean Coulomb, the 'father of geophysics in the Mediterranean'. The newsletter proudly narrated that after four years as director of the Institut de Météorologie et Physique du Globe d'Algérie, Coulomb became head of the Institut de Physique du Globe de Paris in 1941.

Subsequently, he led the Centre National de la Recherche Scientifique (CNRS) and the Centre National d'Etudes Spatiales (CNES) during the 1950s and 1960s. The CRAAG newsletter also mentioned his Grenet–Coulomb seismographs,[115] which were installed in Lebanon's Ksara Observatory, among other places.[116]

Through geodesy and map-making, observatory scientists in collaboration with army officers contributed to the colonisation of Algeria and other countries.[117] However, in celebrating French astronomers as part of the long history of the Algiers Observatory, the CRAAG made little reference to imperialism. The second article on 'figures of the past' in the CRAAG newsletter was dedicated to Charles Trépied, who had become the observatory's director in 1890. The article mentioned Trépied's connections to Paris, where he was born and to whose *Carte du ciel* he contributed in a neutral if not positive tone.[118] The CRAAG praised Jean Lagrula in particular, who had remained in Algeria after its independence. Lagrula was the son of a director of the Algiers Observatory who himself came to lead the Institut de Météorologie et Physique du Globe d'Algérie. A portrait in the centre's newsletter in 2011 introduced Lagrula as one of those 'who through their work leave an indelible mark in the scientific life of a country'. That he had continued his work until 1968 showed 'his deep attachment to Algeria'.[119]

If French settlers could become part of the scientific heritage of an Arab country, it was even more the case with natives. A street in central Cairo near Tahrir Square was named after the 19th-century Mahmoud al-Falaki. A property belonging to the American University in Cairo on this street was thus called the Falaki Campus.[120] Similarly, streets near the American University of Beirut were named after Mansur Jurdak, as well as Bliss, Van Dyck and others. Many residents did not know these streets by their names and navigated the city with the help of landmarks instead. However, at least Google and other information and map services have kept the streets in their memories.[121] Demonstrating Farouk El-Baz's status as a pan-Arab celebrity, a street in an emirate as far away from Egypt as Ras al-Khaimah in the Gulf bore his name.[122] The Syrian cosmonaut Mohammed Fares was the subject of an art exhibit named 'Space Refugee'. Created by the Turkish artist Halil Altındere, the exhibit opened in New York in January 2017.

At the same time, an executive order by President Donald Trump banned Syrian refugees from entering the United States. 'How absurd that he can go to space and not anywhere else,' Altındere said about Fares. 'I decided to pursue it through irony, because irony can't be easily touched by totalitarians.'[123]

Whether in the form of visual artworks or textual accounts, modern Arab astronomical heritage, like its medieval counterpart, could of course become subject to distortion. Orientalist notions of the rise and fall of civilisations shaped narratives of the Arab 'renaissance' (*nahḍah*) of the 19th and early 20th centuries.[124] Journals such as *Al-Muktafaf* and institutions such as the Lee Observatory could thus become objects of nostalgia in a similar way to the writings of Nasir al-Din al-Tusi and the Maraghah Observatory. There is always the risk of underestimating or exaggerating the influence of individuals, whether that of Ibn al-Shatir on Copernicus or that of Farouk El-Baz on the Apollo astronauts. Concepts such as 'Arab' and 'renaissance' are vague and flexible enough to allow for many interpretations. This is perhaps why they are so long-lasting.

However, even a distorted notion of an Arab astronomical renaissance, just like the idea of an Islamic Golden Age, could become an important symbolic resource for future Arab space science. This notion could inspire younger people to follow in the footsteps of Arab astronauts or encourage governments to invest in new observatories. That the region continues to boast many other resources also provides grounds for optimism. These resources include not just energy in the form of fossil fuels and solar power, but also high mountains and clear skies. Proliferating amateur associations in many countries are also indicative of much actual and potential astronomical manpower. Intensifying connections with other parts of the globe could provide expertise and technology not produced locally. The availability of all these resources makes it highly likely that at least some future Arab space science projects will succeed, benefitting not just the region, but the entire world.

# Notes

## PRELIMS

1. Vaz, ‘The Politics’ (2010).
2. Angus Stevenson, *Oxford Dictionary* (2010).
3. University of Chicago Press, *The Chicago Manual* (2010), 660–784.

## CHAPTER 1 THE ARAB CONQUEST OF SPACE

1. Alkhaja, ‘Astronomy’ (2013). Varghese, ‘Qatar’ (2013).
2. Sultan, ‘Glittering prize’ (2013).
3. Arab News, ‘Planet-finder Qatar’ (2010).
4. Yahia, ‘Qatar discovers’ (2012).
5. Valdini, ‘Dr Khalid Al Subai interview’ (2012).
6. Dominik, interview, 7 July 2014.
7. Pollacco et al., ‘WASP-3b’ (2008).
8. Markaz Qaṭar li-‘Ulūm al-Faḍā’ wa-al-Falak, ‘Man naḥnu.’
9. Mujāhid, ‘Al-Shaykh Salmān Āl Thānī’ (2008).
10. Al-Khaṣāwinah, ‘Al-Markaz al-Iqlīmī’ (2014), 18.
11. Kamrava, *Qatar* (2013).
12. E.g., Al-Khalili, *The House of Wisdom* (2011). Saliba, *Islamic Science* (2007).
13. Alsubai et al. ‘Qatar-1b’ (2011). Bryan et al. ‘Qatar-2’ (2012).
14. Al-Subai and Dominik, ‘Conference scientific rationale’ (2013).
15. Al-Khaṣāwinah, ‘Al-Markaz al-Iqlīmī’ (2014).
16. Wakālat Anbā’ al-Imārāt, ‘Ḥākim al-Shāriqah’ (2008).
17. Mahmoud Effendi, ‘Mémoire’ (1858).
18. Mislis et al., ‘SIDRA’ (2016).
19. Mislis et al., ‘The DOHA algorithm’ (2017).
20. Chad, ‘The Forgotten Apogee’ (2013).
21. Stolz, ‘The Lighthouse’ (2013), 53–54.

22. Alsubai et al., 'Qatar-1b' (2011), 716.
23. Bramich et al., 'RR Lyrae Stars in the GCVS' (2014), 11.
24. Wenger et al., 'The SIMBAD astronomical database' (2000), 9.
25. E.g., Saliba, *A History* (1994). King, *In Synchrony* (2005).
26. E.g., Harding, *Space Policy* (2013).
27. Moltz, *Asia's Space Race* (2011).
28. Siddiqi, 'An Asian Space Race' (2010). Lele, *Asian Space Race* (2013).
29. Valdini, 'Dr Khalid Al Subai interview' (2012).
30. Sultan, 'Glittering prize' (2013).
31. Qatar Foundation. 'QF discovers' (2011), 2.
32. Morgan, *Arab Science* (2009), 106.
33. Qatar Foundation. 'QF discovers' (2011), 2.
34. Guessoum, 'Time' (2013), 164.
35. Ibid., 162.
36. Ibid., 161.
37. Al-ʿUlabī, 'Min al-turāth' (2016)
38. Alsubai, interview, 12 March 2015.
39. Filasṭīn al-Yawm, 'Thalāthat "tiliskūbāt"' (2013).
40. D'Afflitto, 'Fantascienza' (1997). Barbaro, *La fantascienza* (2013).
41. Al-Khaṣāwinah, 'Al-Markaz al-Iqlīmī' (2014), 17.
42. Quoted in: Paikowsky and Ben Israel, 'Science' (2009), 1464.
43. Alsubai, interview, 12 March 2015.
44. Flanagan, 'Jordan theme park' (2011).
45. Gonzalez-Quijano, 'Aux frontières' (1997).
46. Kalder, 'Dictator-lit' (2010).
47. Pirard, 'German Rockets' (1997), 895–96. Karp, 'Space technology' (1986), 162–63.
48. Paikowsky and Ben Israel, 'Science' (2009), 1464.
49. Navias, 'Ballistic missile proliferation' (1989): 228.
50. Said, 'Missile proliferation' (2001), 54.
51. Bahgat, 'Proliferation' (2008).
52. Paikowsky and Ben Israel, 'Science' (2009), 1464.
53. Morgan, *Arab Science* (2009), preface.
54. Alsubai, 'Astrophysics & Aerospace Workshop' (2010).
55. Qatar National Research Fund, 'Khalid Al Subai' (2013).
56. Brentjes, 'The Prison' (2012), 132.
57. Saliba, 'The Pebble' (2011).
58. Morgan, *Arab Science* (2009), foreword, preface.
59. Stolz, 'The Lighthouse' (2013).
60. Elshakry, *Reading Darwin* (2013).
61. Sirrs, *Nasser* (2006).
62. Appiah, 'Cosmopolitan Patriots' (1997). Nielsen, 'Cosmopolitan Nationalism' (1999).
63. Freitag, 'Cosmopolitanism' (2010), 1.

## CHAPTER 2 NATIONALISM AND COSMOPOLITANISM

1. ʿUmrān, *Faḍāʾ wāsiʿ* (1997).
2. ʿUmrān, 'Faḍāʾ wāsiʿ' (2014), 113–16.
3. Ibid., 113–18.
4. Schubring, 'Recent Research' (2000).
5. Stolz, 'The Lighthouse' (2013), 127.
6. Asaad and Mikhail, 'Lunar Work' (1979).
7. Dominik, interview, 7 July 2014.
8. M. A. Al Saud et al., 'Broad Band Filter Colours' (1996). M.A. Al-Saud et al., 'Spectroscopic Observations' (1997).
9. Al-Jārūshah, 'Qism al-falak' (2009).
10. New Scientist, 'Reign of terror' (1981).
11. Baars, *International Radio Telescope Projects* (2013), 151–52.
12. Burgess and Vis, *Interkosmos* (2016), 225.
13. Woolley, 'Riyadh Telescope Committee' (1968), 2.
14. Al-Malki, 'Astronomy in Saudi Arabia' (2004), 262.
15. Simmons, 'Seeing Stars in Iraq' (2007), 23.
16. Permanent Delegation of the Arab Republic of Egypt to Unesco, 'Helwan Observatory' (2010).
17. Shears and Shaker, 'Harold Knox-Shaw' (2015), 90–91.
18. Malawi, 'Atomic hydrogen' (1989). Al-Mleaky, 'Density diagnostics' (1990). Basurah, 'Spectropolarimetric studies' (1991).
19. Krige, 'Embedding' (2014).
20. Moulin, 'The French–US space research cooperation' (2003), 645.
21. Sirrs, *Nasser* (2006), 11.
22. Al-Shinnāwī, 'Al-ṣārūkh al-Almānī' (1952), 22.
23. Qunṣul, "Askarat al-fadā"' (2003), 332.
24. Pirard, 'German Rockets' (1997), 886–87.
25. Sirrs, *Nasser* (2006), 82.
26. Bermudez Jr, 'Pyramid scheme' (2010), 48.
27. Chad, 'The Forgotten Apogee' (2013).
28. Al-Bakr, 'Niẓām al-taṭawwuʿ' (1969).
29. United Nations Monitoring, Verification and Inspection Commission, *Compendium* (2007), 345.
30. Ḥusayn, 'Qānūn al-taʿdīl al-rābiʿ' (1988).
31. Harding, *Space Policy* (2013), 143. Day, 'Saddam's space program' (2009).
32. Al-Naʿīmī, 'Al-ʿĀbid al-ʿIrāqī' (1989), 9.
33. Ibid.
34. Al-Naʿīmī, 'Radʿ al-mashrūʿ al-Ṣuhyūnī' (1990).
35. Roberts, 'Paardefontein Satellite Tracking Station' (2012), 165.
36. Irbah et al., 'Astronomy in Algeria' (2001), 172–73. Centre de Recherche en Astronomie, Astrophysique et Géophysique, 'Commémoration' (2010), 6.
37. Sirrs, *Nasser* (2006), 150.

38. Pirard, 'German Rockets' (1997), 887–88.
39. Shapir, 'Satellites' (2004), 26.
40. New Scientist, 'Reign of terror' (1981), 4.
41. Tarikhi, *The Iranian Space Endeavor* (2015), 21, 27.
42. Baars, *International Radio Telescope Projects* (2013), 152.
43. Alnajjar et al., *International Initiative* (2004), 15.
44. Harding, *Space Policy* (2013), 143.
45. Kuwait News Agency, 'Marṣad al-ʿUjayrī al-Falakī' (2002).
46. Al-Malki, 'Astronomy' (1995), 672.
47. Alsabti, 'Astronomy' (2007), 655.
48. Harding, *Space Policy* (2013), 143–44.
49. Said, 'Missile proliferation' (2001), 53.
50. Carus and Bermudez, 'Iraq's Al-Husayn missile programme, part 1' (1990). Windrem, 'Iraqi negotiator' (2003).
51. New York Post, 'Iraq' (2012).
52. Al-Jaffal, 'Iraq's defunct military industry' (2015).
53. Amīn, 'Al-ʿĀbid al-ʿIrāqī' (2011).
54. Makiya, *Republic of Fear* (1998).
55. Ratney, 'Qatari Amir's Wife' (2009).
56. Al-Naimiy, 'Curriculum Vitae' (2010).
57. Al-Sudayrī, 'Al-Prūfīsūr al-Naʿīmī' (2008).
58. Väyrynen, 'The Arab Organization of Industrialization' (1979), 68.
59. Al-Naimiy and Konsul, 'Astronomy' (2001), 70.
60. Qunṣul and al-Naʿīmī, *Taqrīr* (2002), 2.
61. Waṣfī, *Al-insān al-ālī al-qātil* [1990s], 96.
62. ʿUmrān, 'Kutlah ḥayyah' (2010), 28; 'Ruwwād al-Kawkab al-Aḥmar' (2014), 112.
63. Al-Hayʾah al-ʿArabīyah lil-Taṣnīʿ, "ʿAn al-hayʾah' (2017).
64. Qunṣul and al-Naʿīmī, *Taqrīr* (2002), 9, 12, 17.
65. Al-Khaṣāwinah, 'Al-Markaz al-Iqlīmī' (2014), 18.
66. Sardar, 'Middle East brain drain' (1980).
67. Issa and Gamal El Din, 'Astronomical Research Activities' (1986), 92.
68. E.g., Almleaky *et al.*, 'Air Drag Effect' (1996).
69. McDougall, 'Dream' (2011), 253.
70. Guessoum, 'Time' (2013).
71. Qatar Foundation, 'Contents' (2011).
72. ʿUmrān, 'Ruwwād al-Kawkab al-Aḥmar 1/2' (2014), 113.
73. Waṣfī, *Hujūm al-zawāḥif* [1990s], 7–8.
74. Woolley, 'Riyadh Telescope Committee' (1968), 2.
75. McNally, *Proceedings* (1990), 66.
76. Al-Naimiy, 'The Iraqi National Astronomical Observatory' (1986), 51.
77. Al-Naʿīmī, 'Radʿ al-mashrūʿ al-Ṣuhyūnī' (1990), 21.
78. Al-Naimiy and Konsul, 'Basic Space Sciences' (2004), 280.
79. Al-Naimiy, 'The role' (2011), 430.
80. Al-Naimiy and Konsul, 'Astronomy' (2001), 60, 64, 68.

81. Ḥāfiẓ, 'Muʾtamar al-Falak' (2013).
82. Al-Naimiy and Konsul, 'Astronomy' (2001), 70.
83. Arab Union for Astronomy and Space Sciences, *Al-muʾtamar* (2006), 2.
84. Al-Naʿīmī, *Al-kawn wa-asrāruhū* (2000).
85. Guessoum, *Islam's Quantum Question* (2011), 185.
86. E.g., Shaltūt, 'Min al-ishārāt al-kawnīyah' (2004).
87. Al-Saʿdanī, 'Wa-kāna al-insān' (2001).
88. Samaha, 'Notes' (1938).
89. Saliba, 'Solar Observations' (1985), 113.
90. Qunṣul and al-Naʿīmī, *Taqrīr nashāṭāt* (2002), 12.
91. Al-Barghūthī, 'Falakīyan' (2014).
92. Aly and Galal, 'Choice of a Site' (1982), 181. Saʿd, 'Marṣad Jāmiʿat al-Malik ʿAbd al-ʿAzīz' (2011).
93. Woolley, 'Riyadh Telescope Committee' (1968), 2.
94. Al-ʿUmarī, 'Kayfa yuṭlaq' (2014).
95. Abad, interview (2016).
96. Al-Khalīj Ūnlāyin, 'Markaz Qaṭar li-ʿulūm al-falak' (2016).
97. Mimouni, 'Popularising science' (2011), 742. Qassūm et al., 'ʿIlm al-falak fī al-Jazāʾir' (1999).
98. Mimouni, 'Popularising science' (2011), 742.
99. Guessoum, 'Algeria' (1996).
100. Guessoum and Meziane, 'Visibility' (2001), 2.
101. Bū al-Jidarī, 'Al-falakī al-Jazāʾirī' (2015).
102. Watenpaugh, 'Cleansing' (2005).
103. Hanley, 'Grieving Cosmopolitanism' (2008).
104. Al-Shinnāwī, 'Al-ṣārūkh al-Almānī' (1952), 22.
105. Al-Ḥakīm, *Riḥlah ilá al-ghad* (1994), 65–66. See also al-Shārūnī, 'Al-khayāl al-ʿilmī' (2000).
106. Al-Ḥakīm, *Riḥlah ilá al-ghad* (1994), 66.
107. Völkel, *Die Vereinten Nationen* (2008), 357.
108. Al-Fiqī, 'Al-ʿArab wa-al-munaẓẓamāt al-duwalīyah' (2016).
109. Abdel-Rahman, 'Technology' (1967), 114.
110. Appiah, 'Cosmopolitan Patriots' (1997).
111. Al-Fiqī, 'Ibrāhīm Ḥilmī ʿAbd al-Raḥmān' (2009).
112. Marchisio, 'The Evolutionary Stages' (2005), 220.
113. United Nations General Assembly, '1348' (1958).
114. United Nations General Assembly, '1472' (1959).
115. Saad, 'A Writer and Diplomat' (2000).
116. United Nations General Assembly, '1721' (1961).
117. United Nations General Assembly, '32/196' (1977).
118. United Nations General Assembly, '35/16' (1980).
119. COPUOS, 'Verbatim Record' (1974), 21.
120. Abdel-Ghani, 'Man and Space' (1966), 6.
121. United Nations General Assembly, '2222' (1966)
122. Abdel-Ghani, 'The United Nations' (1984), 54.

123. Office of Technology Assessment, *UNISPACE* (1983), 31.
124. COPUOS, 'Report' (1968), 123.
125. Republic of Iraq, *Iraqi Academic Conference* (2009), 38.
126. Schaufele, 'UN Secretariat Staffing' (1974).
127. Bennett, 'USUN Daily Classified Summary' (1974).
128. Habgood, 'Report' (1982), [6].
129. Abdel-Ghani, 'The United Nations and Outer Space' (1984), 58.
130. Office of Technology Assessment, *UNISPACE* (1983), 77.
131. Subcommittee on Space Science and Applications, *Report* (1982), 15.
132. Qadrī, *Hurūb ilá al-faḍā'* (1983), 13–14. See also al-Shārūnī, 'Yūtūbiyā al-khayāl al-ʿilmī fī al-riwāyah al-ʿArabīyah al-muʿāṣirah' (2000).
133. Miller, 'In Saudi Arabia' (1983).
134. Al-Malki, 'Astronomy in Saudi Arabia' (2004), 263.
135. Baars, *International Radio Telescope Projects* (2013), 151.
136. Ibid., 151–52.
137. Alsabti, email to John Hearnshaw (2004).
138. Al-Naimiy and Haubold, 'Preface' (2000), vii.
139. COPUOS, *Report* (1991).
140. Al-Naimy and Konsul, 'Basic Space Sciences' (2004), 286.
141. Haubold, 'UN/ESA workshops' (1998), 111.
142. Shaltout, 'Small Astronomical Telescopes' (2004), 258; 'The Marskhod Egyptian Drill Project' (2004).
143. Al-Naimiy and Haubold, 'Preface' (2000), ix.
144. Tully and Sadsaoud, 'La création' (2003), 'Astronomical Heritage Sites' (2015).
145. Blamont, 'Les premières expériences' (2001). Sirrs, *Nasser* (2006).
146. Neufeld, 'The Nazi aerospace exodus' (2012).
147. Krige, 'Embedding' (2014), 228.
148. Klārk, 'Sa-nusāfir' (1950)
149. ʿAbd al-Wahhāb, *Riḥlah ilá al-Qamar* (1959).
150. Ibid. On the film, see also Westfahl, *The Spacesuit Film* (2012), 270–72.
151. Moulin, 'The French–US space research cooperation' (2003), 645.
152. Kavanaugh, 'Algeria' (1999), 29.
153. Fesenkov, 'Expedition' (1958), 285.
154. Asaad et al., 'A Joint Interkosmos–SAO–HIAG Laser Ranging Project' (1979), 94–95.
155. Al-Naimiy, 'The Iraqi National Astronomical Observatory' (1986), 52.
156. Harding, *Space Policy* (2013), 142, 149–50. Al-Khalīj, 'Al-ʿIrāq' (2016).
157. Asaad et al., 'A Joint Interkosmos–SAO–HIAG Laser Ranging Project' (1979).
158. Shears and Shaker, 'Harold Knox-Shaw' (2015).
159. Republic of Iraq, *Iraqi Academic Conference* (2009), 38.
160. Munns, *A Single Sky* (2012).
161. Kellermann, *David S. Heeschen* (2014), 3.

162. Republic of Iraq, *Iraqi Academic Conference* (2009), 38.
163. Schenectady Gazette, 'Albany State Professor' (1969). George, 'High Resolution Radio Interferometric Observations' (1974), 224.
164. Aiad, 'Teaching' (1990), 398.
165. Müller and Jappel, *Proceedings* (1977), 38.
166. Corbett, *Proceedings* (2010), 417.
167. Alsabti, 'Dr A W Alsabti.'
168. Hearnshaw, 'Commission 46' (2005).
169. McNally, '(10478) Alsabti' (2001).
170. Hearnshaw, 'Commission 46' (2005), 389.
171. Simmons, 'Seeing Stars' (2007), 23.
172. Sargent, 'Itinerant astronomer' (1987).
173. Hamid et al., 'Computation' (1950).
174. Meaburn, 'Zdenek Kopal' (1994): 230.
175. Kopal, 'Preface' (1968), 4.
176. Ibid., 4–5.
177. Samah and Kopal, 'Lunar Photography' (1966). Rifaat, 'Photometric Studies' (1967). Mikhail, 'Color Opposition Effect' (1968).
178. Al-Naimiy, 'Curriculum Vitae' (2010).
179. Kopal, letter to Samaha (10 March 1967), 1.
180. Gerbaldi, 'International Schools' (2007). Kopal, *Of stars* (1986), 263.

## CHAPTER 3 EAST COAST AND NEAR EAST COLLEGES

1. Bingham, 'Solar Eclipse Photography' (1923), 634.
2. David Todd, 'The Singkep Mechanical Commutator' (1902), 170–71. Mabel Todd, *Tripoli* (1912), 127.
3. David Todd, 'Observations' (1906), 463.
4. Mabel Todd, *Tripoli* (1912), 193.
5. Ibid., 3, 50, 122.
6. Ibid., 11, 146–47, 183.
7. David Todd, 'The Singkep Mechanical Commutator' (1902), 170, 174.
8. Mabel Todd, *Tripoli* (1912), 121.
9. Ibid., x.
10. Pang, 'The Social Event of the Season' (1993).
11. Bingham, 'Solar Eclipse Photography' (1923), 633.
12. David Todd, 'The Amherst Eclipse Expedition' (1901), 362.
13. Sheehan and Misch, 'Ménage' (2004), 129.
14. Russell, 'Raymond Smith Dugan' (1940), 466.
15. Penrose, *That They May Have Life* (1941), 74.
16. Mumford, 'Dugan' (2007).
17. Van Aerschodt, 'Raymond Smith Dugan' (1941), 308.
18. Russell, 'Raymond Smith Dugan' (1940), 466.
19. Penrose, *That They May Have Life* (1941), 74.

20. Campbell and Perrine, 'The Lick Observatory–Crocker Eclipse Expedition' (1906), 16–17.
21. Mumford, 'Dugan' (2007), 315.
22. Schayegh, 'Three Questions' (2015), 588. See also Meloy, 'Arab and Middle Eastern Studies' (2016).
23. Anderson, *The American University of Beirut* (2011).
24. Daniel Bliss, *The Reminiscences* (1920), 57.
25. Ibid., 60, 54.
26. David Todd, 'Early History' (1903), 323.
27. Tyler, *History of Amherst College* (1873), 489.
28. Hudson, 'David Todd' (1939), 472–73.
29. Anderson, *The American University of Beirut* (2011), 26.
30. Dorman, 'Missions and Medicine' (2011), 490.
31. Penrose, *That They May Have Life* (1941), 9.
32. Ibid., 14, 17, 27.
33. Abunnasr, 'Another Amherst' (2016), 150.
34. Dorman, 'Missions and Medicine' (2011), 490.
35. Daniel Bliss, *The Reminiscences* (1920), 212–13.
36. Dupont, 'Une école missionnaire et étrangère' (2007).
37. Glaß, *Der* Muqtaṭaf, vol. 1 (2004), 199.
38. David Todd, 'Early History' (1903), 322.
39. Sa'di et al., 'Al-Hakîm' (1937), 25–26.
40. Grafton, *The Contested Origins* (2015).
41. Fandīk, *Kitāb al-mir'āh* (1852).
42. Penrose, *That They May Have Life* (1941), 36.
43. Sa'di et al., 'Al-Hakîm' (1937), 26–27.
44. Van Dyck, 'On the Present Condition' (1849), 568.
45. Makdisi, 'Reclaiming' (1997). Elshakry, 'The Gospel' (2007).
46. Penrose, *That They May Have Life* (1941), 36, 28–29.
47. Vān Dayk, *Uṣūl 'ilm al-hay'ah* (1874), 2.
48. Quoted in Elshakry, 'The Gospel' (2007), 196.
49. Nimr, 'Baʿda ʿahdī bi-ʿilm al-falak' (1938), 484.
50. Todd, 'The Solar Parallax' (1881), 205.
51. Ṣarrūf and Nimr, 'ʿUbūr al-Zahrah' (1882).
52. Misch and Sheehan, 'A movie' (2004), 9.
53. Ṣarrūf and Nimr, 'ʿUbūr al-Zahrah' (1883), 321.
54. Sheehan and Misch, 'Ménage' (2004), 129–30.
55. Misch and Sheehan, 'A movie' (2004).
56. Farag, 'The Lewis Affair' (1972), 73–74.
57. Elshakry, 'The Gospel' (2007), 201.
58. Glaß, *Der* Muqtaṭaf, vol. 1 (2004), 225.
59. Elshakry, 'The Gospel' (2007), 202–03.
60. Ṣarrūf and Nimr, 'ʿUlamāʾ al-hay'ah' (1876).
61. Ṣarrūf and Nimr, 'Muqaddimah' (1876), 1.
62. Ṣarrūf and Nimr, 'Min al-marṣad' (1876), 46.

63. Saliba, 'Copernican Astronomy' (1992), 148–52.
64. Farag, 'The Lewis Affair' (1972), 77–78.
65. Ibid., 79–80.
66. Jeha, *Darwin and the Crisis of 1882* (2004), 126–30.
67. Farah, 'Awakening Interest' (1992), 409.
68. Penrose, *That They May Have Life* (1941), 69.
69. Farag, 'The Lewis Affair' (1972), 82.
70. Journal of the Royal Central Asian Society, 'Obituary' (1952). Glaß, *Der* Muqtaṭaf, vol. 1 (2004), 193.
71. Glaß, *Der* Muqtaṭaf, vol. 1 (2004), 223–31.
72. Nimr, 'Baʿda ʿahdī bi-ʿilm al-falak' (1938), 487.
73. Crossley, 'Percival Lowell' (2000).
74. Ziadat, 'Early Reception' (1994), 20–21.
75. Stolz, 'The Lighthouse' (2013), 228, 230.
76. Elshakry, *Reading Darwin* (2013), 1.
77. American University of Beirut, 'The American University' (2003).
78. Ṣarrūf, 'Qīmat al-baḥth al-ʿilmī' (1926), 71. Abunnasr, 'The Making' (2013), 100.
79. Ṣarrūf, 'Qīmat al-baḥth al-ʿilmī' (1926), 71–72.
80. Ṣarrūf, 'Naḥnu shuʿūb al-ʿālam' (1950).
81. Penrose, *That They May Have Life* (1941), 62, 29.
82. Ibid., 71.
83. West, letter (1885).
84. West, 'Pronunciation of Star Names' (1895).
85. Penrose, *That They May Have Life* (1941), 72, 74.
86. American University of Beirut, 'Lee Observatory.'
87. West, 'A New Variable in *Sculptor*' (1896); 'Five New Southern Variables' (1896); 'Two New Southern Variables' (1896); 'Observations of Southern Variables' (1896); 'A New Variable of Short Period' (1896); 'Confirmations of Variability' (1897); 'Confirmations of Variability' (1898).
88. Carpenter, 'An Astronomical Determination' (1906), 210.
89. Bray, 'A Remarkable Flight' (1895).
90. West, 'Flight of Birds' (1895).
91. Gingerich, 'Five Centuries' (1990), 198.
92. Penrose, *That They May Have Life* (1941), 101.
93. Mahoney, 'Welcome' (2003).
94. Herbig, 'Alfred Harrison Joy' (1974), 526.
95. Campbell and Perrine, 'The Lick Observatory–Crocker Eclipse Expedition' (1906), 13.
96. Ṣarrūf and Nimr, 'Al-kusūf al-muqbil' (1905), 259.
97. Ethel Hussey, 'The Recent Eclipse' (1906), 130, 134.
98. Pang, *Empire* (2002), 38–39.
99. W. J. Hussey, 'The Lick Observatory–Crocker Eclipse Expedition' (1906).
100. Ethel Hussey, 'The Recent Eclipse' (1906), 133.
101. Ṣarrūf and Nimr, 'Al-kusūf' (1905).

102. Pang, *Empire* (2002), 142.
103. Ibid., 33.
104. W. J. Hussey, 'The Lick Observatory–Crocker Eclipse Expedition' (1906), 42.
105. Ibid., 44.
106. Turner, 'From an Oxford Note-Book' (1905), 396.
107. W. J. Hussey, 'The Lick Observatory–Crocker Eclipse Expedition' (1906), 44–45.
108. Penrose, *That They May Have Life* (1941), 122.
109. Joy, 'Transit of Mercury' (1908), 128.
110. Wilson, 'Alfred Harrison Joy' (1975), 226.
111. Joy, 'Photometric measures' (1914).
112. Wilson, 'Alfred Harrison Joy' (1975), 226.
113. Howard S. Bliss, 'The Modern Missionary' (1920), 664.
114. Penrose, *That They May Have Life* (1941), 158–59.
115. Damiens, 'La météorologie' (1995), 29.
116. Berloty, 'L'Observatoire de Ksara' (1912), 105.
117. Udías, *Searching the Heavens* (2003), 176.
118. Damiens, 'La météorologie' (1995), 29.
119. Kobold, 'Mitteilungen' (1918).
120. Anderson, *The American University* (2011), 53.
121. Sayah, 'The American University' (1988), 225.
122. Penrose, *That They May Have Life* (1941), 203.
123. Al-Kulliyeh, 'Two A.U.B. Professors' (1948).
124. Jurdak, *Astronomical Dictionary* (1950), preface.
125. Ibid., preface, y, 81.
126. American University of Beirut, 'Lee Observatory.'
127. Royal Astronomical Society, 'Meeting of the Royal Astronomical Society' (1923), 40–41.
128. National Academy of Sciences, *Report* (1926), 19–20.
129. Gingerich, 'American University Observatory' (1957), 212.
130. Van Maanen et al., 'A Star of Extremely Low Luminosity' (1927), 174.
131. Gingerich, 'American University Observatory' (1957), 212.
132. American University of Beirut, 'Lee Observatory.'
133. Nimr, 'Baʿda ʿahdī bi-ʿilm al-falak' (1938), 487.
134. American University of Beirut, 'Lee Observatory.' Physics Today, 'Julius' (1970).
135. American University of Beirut, 'AUB mourns' (2011).
136. Struve, letter to Jurdak (1954).
137. Oosterhoff, *Ninth General Assembly* (1957), 59.
138. Struve, letter to Jurdak (1955).
139. De Jager and Jappel, *Proceedings* (1971), 5.
140. Rockefeller Foundation, *Annual Report* (1958), 254.
141. Al-ʿAlāwinah, *Dhayl al-aʿlām* (1998), 152.
142. Gingerich and DeVorkin, 'Oral History Transcript' (2005).

143. Gingerich, 'Vita' (2004), 1.
144. Gingerich, 'Eclipse Experiences' (1951); 'Eclipse Saros Series' (1951).
145. Chamberlain and Gingerich, 'Eclipse' (1955).
146. Gingerich, 'A Model' (1959).
147. Gingerich, 'The Naming' (1958).
148. Gingerich, 'Campus Constellations' (2011), 3.
149. Ibid., 6.
150. Gingerich, 'Through Rugged Ways' (1990), 87.
151. Gingerich, 'Vita' (2004), 3.
152. Gingerich, 'PhD Language Requirement' (1977), 9.
153. Gingerich, 'Campus Constellations' (2011), 5.
154. Ibid., 5–6.
155. Saliba, 'AUB's Role' (2016), 218.
156. King, 'Smithsonian Institution Project' (1974).
157. King and Gingerich, 'Some Astronomical Observations' (1982). King, 'The Astronomy' (1983).
158. Morrison, 'IAU Nomenclature' (1976).
159. Millman, 'Working Group' (1977), 351–52.
160. Gingerich, 'Islamic Astronomy' (1986), 83.
161. Gingerich, 'Zoomorphic Astrolabes' (1987).
162. Saliba, 'AUB's Role' (2016), 218.
163. King, 'Frans Bruin' (2002).
164. Driessen, 'Mediterranean Port Cities' (2005), 139.
165. Kennedy et al., 'Edward' (2010), 186.
166. Gingerich, 'Campus Constellations' (2011), 5.
167. Guessoum, 'Time' (2013), 163.
168. American University of Beirut, 'Lee Observatory.'
169. King, 'Frans Bruin' (2002), 214–15.
170. Anderson, *The American University of Beirut* (2011), 188.
171. King, 'Frans Bruin' (2002), 215.
172. Udías, *Searching the Heavens* (2003), 176.
173. Hajjar, 'Astronomy in Lebanon' (2001), 50.
174. Abou Chakra, 'The Lee Observatory' (2013).
175. John Karam, 'Philip Hitti' (2014), 452.
176. Chad, 'The Forgotten Apogee' (2013).
177. Helou, 'Dr. George Helou.'
178. Ibid.
179. Helou, George. 'SIRTF Profiles.'
180. Helou, 'Dr. George Helou.'
181. Emery, 'Choueiri' (2008).
182. Spitzer Science Center, 'Astronomical Success' (2012).
183. El-Dallal, 'The Astronomical Work' (1990), abstract.
184. Dallal, 'Provost Ahmad Dallal.'
185. Cook, 'AUB' (2011).
186. Gingerich, 'Campus Constellations' (2011), 5.

187. Habbal et al., 'Origins' (1997).
188. Glanz, 'Two Spacecraft Track the Solar Wind' (1997), 387.
189. Ilima Loomis, 'This Hawai'i Astronomer' (2015).
190. Joseph, 'Faculty Profile' (2006).
191. United Nations Development Programme, *The Arab Human Development Report* (2006), 108.
192. Altonn, 'UH astronomers' (2006).
193. Habbal et al., 'Localized Enhancements' (2007).
194. Habbal et al. 'Temperature' (2000).
195. Altonn, 'UH astronomers' (2006).
196. St. Cyr, 'Science' (2007), 539.

## CHAPTER 4 STARS OF SCIENCE

1. Qatar is Booming, 'Stars of Science' (2011).
2. Al-Jazīrah, 'Al-jiyūlūjī al-shahīr Fārūq al-Bāz' (2001).
3. Memory Alpha, 'El-Baz'; 'USS Farouk El-Baz.'
4. Farouk El Baz, 'Dr. Farouk El Baz: Public Figure.' Heggy, 'Essam Heggy.'
5. ʿAbd al-Raḥmān, 'ʿAlá hāmish khiṭāb al-ʿarsh' (1947), 1188.
6. Fesenkov, 'The Scientific Work' (1958), 259.
7. Whipple and Hamid, 'On the Origin' (1952). Hamid and Whipple, 'Common Origin' (1963). Hamad, Letter (1964).
8. Hamid and Whipple, 'Tabular Planetary Positions' (1968). Hamid, *First-Order Planetary Theory* (1968).
9. Kissinger, 'Smithsonaian/SFCP' (1973).
10. Krige et al., *NASA* (2013), 3–4.
11. Fu'ād Ṣarrūf, 'Ghazw al-faḍā'' (1970).
12. Al-Sharq al-Awsaṭ, 'Idghār Shuwayrī' (2006).
13. Qadrī, *Hurūb ilá al-faḍā'* (1983), 5.
14. Samaha, letter to Kopal (1967).
15. Mikhail, 'Color Opposition Effect' (1968), 127.
16. Mikhail and Koval, 'Dependence' (1974), 324.
17. Kopal, *Of stars* (1986), 280, 291.
18. Stolz, 'The Lighthouse' (2013), 96–97.
19. Shears and Shaker, 'Harold Knox-Shaw' (2015), 82.
20. Stolz, 'The Lighthouse' (2013), 267.
21. Ibid., 289.
22. Guillermier and Koutchmy, *Total Eclipses* (1999), 110.
23. Madwar, *Preliminary Notice* (1952), 6–11.
24. Madwar, 'Solar Work' (1955), 737. Atkinson, 'Greenwich Eclipse Expedition' (1952).
25. Aly, 'Khartoum Expeditions' (1952).
26. Guillermier and Koutchmy, *Total Eclipses* (1999), 113.
27. Lyot and Dollfus, 'Étude' (1953). Aly, 'Preliminary Note' (1955). Zirin, 'Coronal line' (1955).

28. Aly et al., *A Spectrographic Atlas* (1962); 'A Photometric Study' (1962); 'Emission-Line Ratios' (1963).
29. International Astronomical Union, 'With the Exhibits' (1961).
30. Shaltout, 'Small Astronomical Telescopes' (2004), 252.
31. Samaha, 'The Kottamia 74' Telescope' (1964), 10.
32. Kopal, letter to Clark (1964).
33. Woolley and Aly, 'Radial-Velocity Observations' (1966), 259.
34. Woolley, 'The RR Lyrae Variable Stars' (1965), 130.
35. Woolley and Palmer, 'Halo Characteristics' (1967), 174.
36. Associated Press, 'Star Gazing' (1967).
37. Glass, 'Early Days at SAAO' (2001), 18.
38. Saari and Shorthill, 'Isotherms' (1963).
39. Wilhelms, *To a Rocky Moon* (1993), 91.
40. Saari and Shorthill, 'Thermal Anomalies' (1965). Shorthill and Saari. 'Nonuniform Cooling' (1965). Saari et al., 'Infrared and Visible Images' (1966).
41. Kopal, 'John M. Saari' (1971), 405.
42. Samaha and Kopal, 'Lunar Photography' (1966), 309.
43. Kopal, *Of stars* (1986), 261–62, 276–67.
44. Zakī, 'Wahj al-Qamar' (1959), 78.
45. Kopal, *Of stars* (1986), 273.
46. Meaburn, 'Zdenek Kopal' (1994), 229.
47. Kopal, *Of stars* (1986), 324–25.
48. Samaha, 'Lunar Research' (1967), 87.
49. North, *Observing the Moon* (2000), 70.
50. Brenchley, *Britain* (2005), 65.
51. Kerr, 'Egyptian Foreign Policy' (1968), 117.
52. Kopal, letter to Asaad (1968).
53. Kopal, *Of stars* (1986), ix.
54. Kopal, letter to Samaha (21 December 1967), 1.
55. Kopal, *Of stars* (1986), 280.
56. Meaburn, 'Zdenek Kopal' (1994), 229.
57. Kopal, 'John M. Saari' (1971), 406–07. On these hot spots, see also Winter, 'The infrared moon' (1970).
58. Peacock, 'Multicolor Photoelectric Photometry' (1968).
59. Mikhail, 'Color Opposition Effect' (1968).
60. Asaad and Mikhail, 'Colour Contrasts' (1974), 298.
61. Mikhail, 'Telescopic Investigation' (1979), 209.
62. Al-Jazīrah, 'Al-jiyūlūjī al-shahīr' (2001).
63. Bates, 'Space-Age Immigrant' (1976), 21. Bass, *Reinventing* (1994), 147.
64. Ḥusnī, 'Fī dā'irat al-ḍaw'' (2001).
65. Bass, *Reinventing* (1994), 147–48.
66. Ibid., 148.
67. Bates, 'Space-Age Immigrant' (1976).
68. Pat El-Baz, 'Pat O'Leary El-Baz Writes.'

69. El-Baz and Wright, 'Farouk El-Baz' (2009). Bass, *Reinventing* (1994), 149.
70. Bass, *Reinventing* (1994), 149.
71. Farouk El-Baz, *Say It in Arabic* (1968).
72. Al-Jazīrah, 'Al-jiyūlūjī al-shahīr' (2001).
73. El-Baz and Wright, 'Farouk El-Baz' (2009).
74. Ibid.
75. Silbermann, 'Historical Note' (2001).
76. El-Baz and Wright, 'Farouk El-Baz' (2009).
77. Ibid.
78. Wilhelms, *To a Rocky Moon* (1993), 161.
79. El-Baz and Wright, 'Farouk El-Baz' (2009).
80. Farouk El-Baz, 'Training *Apollo* astronauts' (2011), 52–53.
81. El-Baz and Wright, 'Farouk El-Baz' (2009).
82. Osama, 'Profiles' (2001).
83. Farouk El-Baz, 'Training *Apollo* astronauts' (2011), 53–54.
84. Ibid., 56. El-Baz and Wright, 'Farouk El-Baz' (2009).
85. Bates, 'Space-Age Immigrant' (1976).
86. Bass, *Reinventing the Future* (1994), 132.
87. Farouk El-Baz, 'Training *Apollo* astronauts' (2011), 58.
88. Bass, *Reinventing the Future* (1994), 132. Worden, *Hello Earth* (1974), 48–49.
89. Worden, *Hello Earth* (1974), 34.
90. Ibid., 71, 74.
91. Farouk El-Baz et al., 'Astronaut Observations' (1972).
92. El-Baz and Roosa, 'Significant results' (1972).
93. El-Baz and Wright, 'Farouk El-Baz' (2009).
94. Sagdeev and Eisenhower, 'United States–Soviet Space Cooperation' (2008).
95. El-Baz and Wright, 'Farouk El-Baz' (2009).
96. Bates, 'Space-Age Immigrant' (1976).
97. Farouk El-Baz, 'Discovery' (1972), 38.
98. Farouk El-Baz, 'Al-Khwarizmi' (1973), 1176.
99. Millman, 'Working Group' (1977), 325–42.
100. El-Baz and Gheith, 'Memorial of Riad Abdel-Megid Higazy' (1970).
101. Sagdeev and Eisenhower, 'United States–Soviet Space Cooperation' (2008).
102. Al-Bāz, 'Taqdīm' (1975), 10.
103. Paganelli, 'Visit' (1974).
104. Paganelli, 'Smithsonian Activities' (1975).
105. Paganelli, 'Visit' (1974)
106. Paganelli, 'Smithsonian Activities' (1975).
107. Al-Bāz, 'Dirāsat al-Arḍ' (1976), 87.
108. Osama, 'Profiles' (2001).
109. Farouk El-Baz, 'Expanding desert' (1977), 36.
110. Farouk El-Baz, 'Training *Apollo* astronauts' (2011), 63.

111. Kissinger, 'Visit' (1976).
112. Naifeh, 'Astronauts Visit' (1976).
113. Farouk El-Baz, 'Zhao's Impact' (1998), 468.
114. Farouk El-Baz, 'Journey' (1980), 62.
115. Farouk El-Baz, 'Expanding desert' (1977), 36–38.
116. Khoury, 'Osama Al Baz' (2013).
117. Farouk El-Baz, 'Narrative of the Journey' (1980).
118. Farouk El-Baz, 'Dr. Farouk El-Baz, Director.'
119. Al-Bāz, 'Riḥalāt al-faḍā'' (1978), 30.
120. Al-Bāz, 'Taqdīm' (1997), 9. De Vaucouleurs et al., 'The New Martian Nomenclature' (1975), 91.
121. Al-Maqrīzī, *Towards a Shi'i Mediterranean Empire* (2009), 78.
122. Farouk El-Baz, 'Expanding desert' (1977), 40.
123. El-Baz and Maxwell, *Desert Landforms* (1982).
124. El-Baz and Wright, 'Farouk El-Baz' (2009).
125. Farouk El-Baz, 'Earth Orbital Photography' (1978), 71.
126. Wilhelms, *To a Rocky Moon* (1993), 349.
127. Farouk El-Baz, 'Interdisciplinary Research' (1989), 1132.
128. Bass, *Reinventing* (1994), 140–41.
129. Appiah, 'Cosmopolitan Patriots' (1997).
130. Bass, *Reinventing* (1994), 149.
131. Ibid., 141.
132. Farouk El-Baz, 'Do people make deserts?' (1990).
133. Al-Ḥarīrī, '18 ʿāliman ʿarabīyan' (1983), 86, 89.
134. McCray, *The Visioneers* (2013).
135. Farouk El-Baz, 'Development Corridor.'
136. Osama, 'Profiles' (2001).
137. Choudry, 'Interview with Prof Farouk El-Baz' (2012).
138. Farouk El-Baz, 'Development Corridor.'
139. Khoury, 'Osama Al Baz' (2013).
140. Fahmy, 'Pioneers' (2005).
141. Osama, 'Profiles' (2001).
142. Choudry, 'Interview with Prof Farouk El-Baz' (2012).
143. Al-Jazīrah, 'Al-jiyūlūjī al-shahīr' (2001).
144. Osama, 'Profiles' (2001).
145. Abūqurayn, 'Mamarr (al-Tanmiyah wa-al-Taʿmīr)' (2011).
146. Al-Saʿīd, 'Mamarr al-Tanmiyah' (2011).
147. Al-Ḥarīrī, '18 ʿāliman ʿarabīyan' (1983), 89.
148. Al-Shamlan, 'Introduction' (2000).
149. RAND, 'Farouk El-Baz' (2004).
150. Culbertson et al., *Launching* (2012), iii.
151. Geological Society of America, 'Farouk El-Baz Award' (2013).
152. El-Baz, 'GSA Student Award.'
153. Farouk El-Baz, '1001 Wells.'
154. Plackett, 'A Conversation' (2014).

155. Mīlād, 'Shāhid' (2016).
156. Heggy, 'Etude' (2002), 8.
157. Al-Maṣrī, ''Iṣām Ḥajjī' (2009).
158. Zohny, 'NASA maps' (2012).
159. Heggy, 'Etude' (2002), 3.
160. O'Neill, 'How Mursi's English' (2013). Behar, 'There He Goes Again' (2013).
161. Saied, 'Is there hope' (2015).
162. Cable News Network, 'Ḥajjī ya'tadhir' (2016).
163. Badawī, ''Iṣām Ḥajjī fī ḥiwār' (2015).
164. Ahram Online, 'Claims of Cure' (2014).
165. Jones, 'Top Egyptian Comedian' (2014).
166. Badawī, ''Iṣām Ḥajjī fī ḥiwār' (2015).
167. Zol-Faqar, 'President's scientific adviser' (2014).
168. Saied, 'Is there' (2015).
169. Bohannon, 'A challenge' (2014).
170. Shiha, 'A Novel Method' (2013).
171. Yahia, 'The False Science' (2014).
172. Raydan, 'No cure' (2014).
173. Al-Najjār, ''Afwan 'Iṣām Ḥajjī' (2014).
174. Ahram Online, 'Alexandria Uni' (2015).
175. Badawī, ''Iṣām Ḥajjī fī ḥiwār' (2015). El-Behary, 'Omar Samra' (2013).
176. Saied, 'Is there' (2015).

## CHAPTER 5 SPACE PIONEERS

1. Ibn Baz, *Al-adillah* (1982), 21. Ende, 'Religion, Politik und Literatur' (1982), 381–82. Ibn Bāz, 'Takdhīb wa-naqd.'
2. Yamānī, *Naẓarāt 'ilmīyah* [c. 1978], 6–9.
3. Ibn Bāz, 'Radd 'alá al-muftarīn.'
4. Ibn Baz, *Al-adillah* (1982).
5. Lacey, *Inside the Kingdom* (2009), 90.
6. Lawton and Moody, 'A Prince in Space' (1986).
7. Abdel Haleem, *The Qur'an* (2005), 354. Akel Kahera, personal communication, 15 October 2016.
8. Fabian and Ross-Nazzal, 'Oral History Transcript' (2006), 56.
9. McCray, *The Visioneers* (2013), 140.
10. Mullane, *Riding Rockets* (2006), 197.
11. Hanania, 'Christa McAuliffe' (2016).
12. Muftāḥ, *Al-samā'* (1985), 115. 362.
13. Lewis, 'Muslims in Space' (2013), 111.
14. Ibn Bāz, 'Ḥukm al-qawl.'
15. Rockwell, 'Space Propaganda' (2012).
16. 'Abd al-Raḥmān, 'Al-qamar al-ṣināʿī' (1957), 11, 15.

17. Al-Ḥuṣrī, 'Ittaḥadnā' (1959), 20.
18. Nu'mān, 'Muṣṭalaḥāt' (1959), 362.
19. Goldstein, 'INTELSAT' (1976), 742.
20. Parks, *Cultures in Orbit* (2005), 26
21. Doyle, 'Communication Satellites' (1967), 436.
22. Parks, *Cultures in Orbit* (2005), 21–22, 27.
23. Abu-Argoub, 'Historical, Political And Technical Development' (1988), 62–63, 68.
24. El-Baz and Mitchell, 'Remote Sensing' (1975), 10.
25. Kraidy and Khalil. *Arab Television Industries* (2009), 17.
26. Von Kries, 'The ARABSAT Agreement' (1978), 195.
27. Abu-Argoub, 'Historical, Political And Technical Development' (1988), 75–85.
28. Clark, 'Arabsat' (1985).
29. Wheatley, 'The Planning and Commissioning' (1970), 242–43.
30. Scrafford, 'Communications Explosion' (1976), 692–93.
31. Von Kries, 'The ARABSAT Agreement' (1978), 194–95.
32. Durra and Christie, 'Arabsat' (1985).
33. Al-Mashat, 'The Arab Satellite Communications System' (1982), 187.
34. Yamānī, *Aqmār ṣināʿīyah* (1984), 68, 70.
35. Christie and Moody, 'Arabsat' (1985).
36. Ibid.
37. Yamānī, 'Aqmār al-faḍā'' (1984), 9, 13.
38. Yamānī, *Aqmār ṣināʿīyah* (1984), 20–21.
39. Clark, 'Arabsat' (1985).
40. Abu-Argoub, 'Historical, Political And Technical Development' (1988), 76, 86, 110.
41. Walgate, 'French to make Arabsat' (1981), 372.
42. Sultan Al Saud et al., *One Planet* (2011), 17.
43. Ibid., 17–18.
44. Abu-Argoub, 'Historical, Political And Technical Development' (1988), 87.
45. Sultan Al Saud et al., *One Planet* (2011), 22.
46. Durra and Christie, 'Arabsat' (1985).
47. Evans, *Tragedy* (2012), 352.
48. Harding, *Space Policy* (2013), 75.
49. Yamānī, *Aqmār ṣināʿīyah* (1984), 8.
50. Abu-Argoub, 'Historical, Political And Technical Development' (1988), 110–23.
51. Sultan Al Saud et al., *One Planet* (2011), 24.
52. Broad, 'Shuttle' (1985).
53. Sultan Al Saud et al., *One Planet* (2011), 31–32.
54. Ibid., 42.
55. Creighton and Ross-Nazzal, 'Oral History Transcript' (2004), 34.
56. Sultan Al Saud et al., *One Planet* (2011), 42.

57. Ibid., 36–37.
58. Evans, *Tragedy* (2012), 353.
59. Sultan Al Saud et al., *One Planet* (2011), 282.
60. Evans, *Tragedy* (2012), 351.
61. Ibid.
62. Fabian and Ross-Nazzal, 'Oral History Transcript' (2006), 51.
63. Evans, *Tragedy* (2012), 352.
64. Creighton and Ross-Nazzal, 'Oral History Transcript' (2004), 34.
65. Brandenstein and Butler, 'Oral History Transcript' (1999), 28.
66. Evans, *Tragedy* (2012), 353.
67. Brandenstein and Butler, 'Oral History Transcript' (1999), 28.
68. Sultan Al Saud et al., *One Planet* (2011), 253.
69. Nagel and Ross-Nazzal, 'Oral History Transcript' (2002), 17–18.
70. Sultan Al Saud et al., *One Planet* (2011), 174.
71. Ibid., 176.
72. Fabian and Ross-Nazzal, 'Oral History Transcript' (2006), 54.
73. Evans, *Tragedy* (2012), 363.
74. Fabian and Ross-Nazzal, 'Oral History Transcript' (2006), 54.
75. Abu-Argoub, 'Historical, Political And Technical Development' (1988), 125.
76. Broad, 'Shuttle' (1985).
77. Ulmer et al., 'Spartan 1 X-Ray Observations' (1987).
78. Harland, *The Story of the Space Shuttle* (2004), 95.
79. Sultan Al Saud et al., *One Planet* (2011), 63–66.
80. Lawton and Moody, 'A Prince in Space' (1986).
81. Evans, *Tragedy* (2012), 354.
82. Ibid.
83. Storey and Fitchett, 'Down to Earth' (1986).
84. Evans, *Tragedy* (2012), 355.
85. Sultan Al Saud et al., *One Planet* (2011), 269.
86. Ibid., 202.
87. Storey and Fitchett, 'Down to Earth' (1986).
88. Sultan Al Saud et al., *One Planet* (2011), 198, 219.
89. Lawton and Moody, 'A Prince in Space' (1986).
90. Al-Saud and Chang, 'Space Shuttle RCS Plumes' (1988). Turki Al-Saud, 'Three-Dimensional Computerized Gas Emission Tomography' (1997), 9.
91. El-Faer et al., 'Phase Separation in Microgravity' (1988).
92. Al-Rajeh, 'Sensorimotor Adaptation' (1988). Viéville et al., 'Adaptive Modifications' (1986). Clément et al., 'Modifications' (1986). Clément and Lestienne, 'Adaptive modifications' (1988). Clément and Berthoz, 'Cross-Coupling' (1990).
93. Sultan Al Saud et al., *One Planet* (2011), 246.
94. Storey and Fitchett, 'Down to Earth' (1986).
95. Sultan Al Saud et al., *One Planet* (2011), 280.
96. Ibid., 281–82.

97. Nagel and Ross-Nazzal, 'Oral History Transcript' (2002), 26–27. Evans, *Tragedy* (2012), 363.
98. Fabian and Ross-Nazzal, 'Oral History Transcript' (2006), 59–60. Evans, *Tragedy* (2012), 363.
99. Fabian and Ross-Nazzal, 'Oral History Transcript' (2006), 60. Mullane, *Riding Rockets* (2006), 42. Evans, *Tragedy and Triumph in Orbit* (2012), 364.
100. Storey and Fitchett, 'Down to Earth' (1986), 33.
101. Ibid., 36.
102. Evans, *Tragedy* (2012), 364. Fabian and Ross-Nazzal, 'Oral History Transcript' (2006), 60–61.
103. Storey and Fitchett, 'Down to Earth' (1986).
104. Storey, 'Question Time' (1986), 35.
105. Lawton and Moody, 'A Prince in Space' (1986), 29.
106. Association of Space Explorers, 'I Congress' (1985).
107. Association of Space Explorers. 'V Congress' (1989).
108. Association of Space Explorers, 'Statement' (1989).
109. Storey, 'Question Time' (1986).
110. Lawton and Moody, 'A Prince in Space' (1986).
111. Sultan Al Saud et al., *One Planet* (2011), 306. Al-Rashīd, 'Awwal rāʾid faḍāʾ ʿArabī' (2012).
112. Abu-Argoub, 'Historical, Political And Technical Development' (1988), 80–81, 85, 91.
113. Kandil, 'Le satellite d'Aladin' (1987), 667.
114. Fawzī, 'Al-qamar al-ṣināʿi al-ʿArabī' (1986).
115. Gau and Sourbes, 'Un système régional' (1989), 20.
116. Sakr, *Satellite Realms* (2001), 10–11. Kraidy and Khalil. *Arab Television Industries* (2009), 17.
117. Harland, *The Story of the Space Shuttle* (2004), 71–72.
118. Karimi Alavi, 'Contradictions' (2010), 189.
119. Telhami, 'Power' (1999).
120. Rinnawi, *Instant Nationalism* (2006).
121. Phillips, '*Team Arab*' (2012).
122. Al-Hinai et al., 'Shuttle Imaging Radar Views' (1997). Dabbagh et al., 'Detection' (1997).
123. Sultan Al Saud, 'Biography' (2013).
124. Danforth, *Crossing the Kingdom* (2016), 155–56.
125. Sultan Al Saud et al., *One Planet* (2011), 300–01.
126. Ibid., 550.
127. Ibrahim and Koudmani, 'A Brief Account' (1998), 345.
128. Harding, *Space Policy* (2013), 171.
129. Sardar, 'Shake-up' (1980).
130. Saliba, 'AUB's Role' (2016), 219.
131. Ibrahim and Koudmani, 'A Brief Account' (1998), 346.
132. Koudmani, 'A Brief Account' (1992), 158.
133. Al-Aḥmad, 'Rāʾid al-faḍāʾ Muḥammad Fāris' (2007).

134. Garthwaite, 'From astronaut to refugee' (2016).
135. Burgess and Vis, *Interkosmos* (2016), 229.
136. ʿAṭīyah, '"Al-Dustūr" tuḥāwir thānī rāʾid faḍā'' (2009).
137. Fares and Edelby, 'TM-3' (2005), 226.
138. Harland, *The Story of Space Station Mir* (2005), 160–61, 306.
139. Baker, *The Story of Manned Space Stations* (2007), 83–84.
140. Burgess and Vis, *Interkosmos* (2016), 225.
141. Quoted in: Nizameddin, *Russia* (1999), 53.
142. Fares and Edelby, 'TM-3' (2005), 227.
143. Al-Daybān, '(Muḥammad Fāris)' (2014).
144. Ghazal, 'The first Syrian' (2015).
145. Hussein Ibrahim et al., 'GORS' (2000), 146–47.
146. Harland, *The Story of Space Station Mir* (2005), 161.
147. Fares and Edelby, 'TM-3' (2005), 229.
148. True, 'Rāʾid al-faḍāʾ Muḥammad Fāris' (2011).
149. Ghazal, 'The first Syrian' (2015).
150. Ibid.
151. Burgess and Vis, *Interkosmos* (2016), 234.
152. Kelley, *The Home Planet* (1988), 77.
153. Burgess and Vis, *Interkosmos* (2016), 234.
154. Al-Daybān, '(Muḥammad Fāris)' (2014). Al Arabiya, 'Rāʾid faḍāʾ sūrī' (2013). Sharma, 'Syrian in space' (2015).
155. Garthwaite, 'From astronaut to refugee' (2016).
156. Al-Daybān, '(Muḥammad Fāris)' (2014).
157. Āl Muḥammad, 'Muqābalah maʿa rāʾid al-faḍāʾ.'
158. Al-Aḥmad, 'Rāʾid al-faḍāʾ Muḥammad Fāris' (2007).
159. Āl Muḥammad, 'Muqābalah maʿa rāʾid al-faḍāʾ.'
160. Al-Aḥmad, 'Rāʾid al-faḍāʾ Muḥammad Fāris' (2007).
161. Okaz, 'Awwal rāʾid faḍāʾ Sūrī' (2012).
162. Al-Ghaḍawī, 'Al-niẓām' (2012).
163. Hijris, 'Rāʾid al-faḍāʾ "al-Fāris"' (2013).
164. Shahdā, 'Al-Suwayd' (2015).
165. Sharma, 'Syrian in space' (2015).
166. Al-Ghaḍawī, 'Al-niẓām' (2012).
167. Zambrana, 'AP Interview' (2016).
168. Ghazal, 'The first Syrian' (2015).

## CHAPTER 6 THE ARABS' FINAL FRONTIER

1. Bhattacharjee, 'Saudi Universities' (2011).
2. Hare, 'Saudis recruit' (2012).
3. Al-Khedairy, 'Saudi University Policy' (2012).
4. Al-Riyāḍ, 'Awāmir malakīyah' (2012).
5. Al-Muqrin, 'Li-mādhā khasarnā al-ʿUthmān?' (2012).

6. Miley, 'Saudi University Policy' (2012), 1041.
7. Hare, 'Saudis recruit' (2012).
8. Bhattacharjee, 'Saudi Universities' (2011).
9. Bornmann and Bauer, 'Which of the World's Institutions' (2015), 2147.
10. King Abdulaziz University, 'Former Distinguished Adjunct Professors' (2015).
11. Zahed, 'Saudi University Policy' (2012), 1040.
12. Ibid.
13. Miley, 'Saudi University Policy' (2012), 1041.
14. Becker, 'Saudi University Policy' (2012).
15. Al-Luḥaydān, 'Ahammīyat inshā' wakālat al-faḍā' al-Su'ūdīyah' (2009).
16. Ibid.
17. Al-Luḥaydān, 'Ahammīyat inshā' wakālat al-faḍā' al-Su'ūdīyah (2)' (2009).
18. Ibid.
19. Satia, 'Drones' (2014).
20. Abad, 'Man to Mars' (2011).
21. Al-Qar'ān et al., ''Alá gharār mashrū' "Abūlū"' (2014), 12, 15.
22. Alfaisal University, 'Alfaisal faculty' (2014)
23. Abad, 'Faisal AlMubarak Unveils' (2015); 'Prologue' (2015).
24. Abad, 'When Dr. Faisal AlMubarak Fired me' (2015).
25. Shapir, 'Satellites ' (2004), 25–28.
26. Harding, *Space Policy* (2013), 167.
27. Aman, 'Egyptian Satellite' (2014).
28. Ibid., 28.
29. Es'hailSat, 'Timeline' (2016). On Canopus, see also Bailey, 'Bedouin Star-Lore' (1974). Varisco, 'Folk Astronomy' (1990).
30. Kass, 'Iran's Space Program' (2006), 26.
31. Shapir, 'Satellites' (2004), 27–28.
32. Harding, *Space Policy* (2013), 167.
33. Zeinobia, 'Lost' (2010).
34. Cairo Post, 'Communication' (2015).
35. Space Daily, 'Egypt Mulls' (2015)
36. SpaceWatch Middle East, 'China' (2017).
37. Navias, 'Ballistic missile proliferation' (1989), 229.
38. Senn, 'The Arms-Dynamic Pacemaker' (2009), 56–57.
39. Tarikhi, 'Iran's space programme' (2009).
40. Inhorn, *Cosmopolitan Conceptions* (2015).
41. Masad, 'Dubai' (2008).
42. Inhorn, 'Cosmopolitan conceptions in global Dubai?' (2016), 25.
43. Orton, 'The desert of the unreal' (2012).
44. Elsheshtawy, 'The Prophecy' (2011).
45. Damon and Lister, 'Tunisian 'Star Wars' (2011).
46. Riley, 'Star Wars' (2015).
47. Johnson, 'Science fiction' (2012).

48. Ahmad, 'Interview' (2012).
49. Jalal, 'These Guys' (2016).
50. MacLean, 'Time' (2016), 197.
51. Bailey, 'Bedouin Star-Lore' (1974), 585.
52. Shapir, 'Satellites' (2004), 26.
53. Lele, *Asian Space Race* (2013), 38.
54. Ibid. Schreck, 'Abu Dhabi' (2009).
55. Al-Shaykh, 'Awwal Imārātī' (2007).
56. Croucher, 'Work to start' (2014).
57. Shaaban, 'One-way trip to Mars' (2014).
58. Qubaysī, 'Miṣrī ikhtārūhu' (2015). Aboulkheir, 'Meet Mohammad Sallam' (2015).
59. Farouk El-Baz, *The Arab World* (1998).
60. Wakālat Anbāʾ al-Imārāt, 'Ḥākim al-Shāriqah' (2008).
61. Khaleej Times, 'Khalifa Sat' (2013).
62. Lobel, 'The United Arab Emirates reaches' (2015).
63. Al Qasim et al., 'Nayif-1' (2016), 1.
64. Bassi, *A Scientific Peak* (2015).
65. Alrais et al., 'Emirates Mars Mission' (2016), 2.
66. Alrais et al., 'Emirates Mars Mission' (2016).
67. Gulf News, 'UAE unveils details' (2015).
68. Malek, 'UAE's Mars space mission' (2015).
69. Waṣfī, *Al-ḥubb al-mustaḥīl* [1990s], 152–53.
70. Gulf News, 'UAE unveils details' (2015).
71. Shaheen, 'Emirates Space Mission' (2015).
72. Gulf News, 'UAE unveils details' (2015).
73. Shaheen, 'Emirates Space Mission' (2015).
74. Gulf News, 'UAE to build first city' (2017).
75. Al-Bayān, 'Ruwwād faḍāʾ al-Imārāt' (2017).
76. SpaceWatch Middle East, 'UAE's MBRSC' (2017)
77. Hany Rashwan, personal communication.
78. Egyptian Streets, 'Egypt' (2017).
79. Guessoum, 'Arab Space Programs' (2014).
80. Cable News Network, 'Al-Jazāʾir wa-al-Imārāt' (2017).
81. Querci and Querci, 'High Precision Photometry' (1992).
82. Querci et al., 'A complementary network' (1995), 85–86.
83. Querci et al., 'Network of Oriental Robotic Telescopes' (1995), 239.
84. Ibid.
85. Querci, 'Network of Oriental Robotic Telescopes' (1995).
86. Hajjar et al., 'The Network of Oriental Robotic Telescopes (NORT)' (2000), 45, 47.
87. Al-Mousli, 'Astronomical Site Testing' (2001).
88. Farooqui et al., 'Establishment of an Observatory' (2002).
89. Sultan and Querci, 'The Mountains of Yemen' (2002), 72–73.

90. Centre de Recherche en Astronomie, Astrophysique et Géophysique, 'Visite' (2003).
91. Mimouni, 'Popularising science' (2011), 741–42.
92. Mimouni and Seghouani, 'L'Ecole Doctorale' (2008), 5.
93. Guessoum, 'The future' (2014).
94. Sciences et avenir, 'Le colonel Kadhafi' (2005).
95. Masevich and Hamal, 'Intercosmos Laser Ranging Stations' (1977).
96. Makram Ibrahim et al., 'Satellite Laser Ranging' (2015), 126
97. Novotny and Blazej, 'Human factor' (2007).
98. Bachelet et al. 'MOA 2010-BLG-477Lb' (2012).
99. El-Awady, 'Qatar discovers' (2016). Alsubai et al., 'Qatar Exoplanet Survey' (2017).
100. Hill et al., 'The Global Oscillation Network Group Site Survey' (1994).
101. Gillon, 'Seven temperate terrestrial planets' (2017), 456.
102. Laylin, 'NASA Made a Crazy Planet Discovery' (2017).
103. Gillon, 'Seven temperate terrestrial planets' (2017).
104. Laylin, 'NASA Made a Crazy Planet Discovery' (2017).
105. Hiddleston, 'Nearby star' (2017).
106. Cordall, 'Star Wars' abandoned Tunisian locations' (2015).
107. Azoulay, 'Diversité' (2006).
108. Gorman, 'The archaeology' (2009), 132.
109. Tully et al., 'Les traces' (2008), 216.
110. Tully et al., 'Astronomical Heritage Sites' (2015).
111. Permanent Delegation, 'Helwan Observatory' (2010).
112. Abou Chakra, 'The Lee Observatory' (2013).
113. Issa, 'Petition Circulates' (2016).
114. Austen, 'To Infinity' (2011).
115. Centre de Recherche en Astronomie, Astrophysique et Géophysique, 'Jean Coulomb' (2004).
116. Udías, *Searching the Heavens* (2003), 177.
117. Schiavon, 'Geodesy and Map-Making' (2010).
118. Centre de Recherche en Astronomie, Astrophysique et Géophysique, 'Charles Trépied' (2003).
119. Centre de Recherche en Astronomie, Astrophysique et Géophysique, 'Jean Lagrula' (2011).
120. Muṣṭafá, 'Tārīkh ashhar aḥyāʾ al-Qāhirah' (2015).
121. Michael Karam, 'Orderly fashion' (2015).
122. Farouk El-Baz, 'A Street in Ras Al-Khaima.'
123. Selin Thomas, 'Dark Reality' (2017).
124. El-Ariss, 'Let There Be *Nahdah*!' (2015).

# BIBLIOGRAPHY

## INTERVIEWS

Hamid Al-Naimiy, via telephone, 28 December 2016.
Hisham Abad, via Twitter, March 2016.
Ivan Procházka, via telephone, 21 March 2016.
Keith Horne, St Andrews, 3 July 2014.
Khalid Alsubai, Doha, 9 September 2013 and 12 March 2015.
Martin Dominik, St Andrews, 7 July 2014.
Nidhal Guessoum, via Skype, 4 January 2017.
Sonya Shaykhoun, Doha, 12 January 2016.

## PRIMARY SOURCES


Abad, Hisham. 'Man to Mars.' *YouTube*, 12 April 2011. https://www.youtube.com/watch?v=cDoIoNenzL8.

———. 'Prologue.' *Alfaisal Veterans*, August 2015. http://alfaisal-university.blogspot.com/2015/08/prologue.html.

———. 'Faisal AlMubarak Unveils his Corruption.' *Alfaisal Veterans*, 9 August 2015. http://alfaisal-university.blogspot.com/2015/11/faisal-almubarak-reveals-his-corruption.html.

———. 'When Dr. Faisal AlMubarak Fired me from Alfaisal University.' *Alfaisal Veterans*, 7 August 2015. http://alfaisal-university.blogspot.com/2016/01/when-dr-faisal-almubarak-fired-me-from.html.

ʿAbd al-Raḥmān, Ibrāhīm Ḥilmī [see also Abdel-Rahman, Ibrahim Helmi]. 'ʿAlá hāmish khiṭāb al-ʿarsh: al-ʿilm gharīb fī bilādinā.' *Al-Thaqāfah*, no. 465 (25 November 1947): 1188–92.

———. 'Al-qamar al-ṣināʿī wa-mā ṣanaʿ.' *Al-Majallah*, no. 11 (November 1957): 10–15.

ʿAbd al-Wahhāb, Ḥamādah. *Riḥlah ilá al-Qamar*. Behna Films, 1959.

Abdel-Ghani, A.H. 'Man and Space: A Great International Venture.' *UNESCO Courier* 19 (May 1966): 5–6.

———. 'The United Nations and Outer Space: Some Observations.' In *Maintaining Outer Space for Peaceful Uses*, edited by Nandasiri Jasentuliyana. Tokyo: United Nations University, 1984.

Abdel Haleem, M.A.S. *The Qur'an: A new translation*. Oxford: Oxford University Press, 2005.

Abdel-Rahman, Ibrahim Helmi [see also 'Abd al-Raḥmān, Ibrāhīm Ḥilmī]. 'Technology and the Developing Countries.' In *Technology and World Trade: Proceedings of a Symposium*, edited by Robert Stern, 113–18. Washington: National Bureau of Standards, 1967.

Abou Chakra, David. 'The Lee Observatory: A Staple of AUB Left to Rust.' *Beirut News Network*, 2 January 2013. http://beirutnewsnetwork.com/david/2013/01/02/the-lee-observatory-a-staple-of-aub-left-to-rust-final/.

Aboulkheir, Rajia. 'Meet Mohammad Sallam: Egypt's potential Mars settler.' *Al Arabiya*, 25 February 2015. http://ara.tv/4byvm.

Abūqurayn, ʿAntar ʿAbd al-ʿĀl. 'Mamarr (al-Tanmiyah wa-al-Taʿmīr). ru'yah naqdīyah.' *Al-Shurūq*, 14 April 2011. http://www.shorouknews.com/columns/view.aspx?cdate=14042011&id=7e382c77-fc0e-4cb7-9ca4-7298414b0f6f.

Ahmad, Muhammad Aurangzeb. 'Interview of Noura Al Noman.' *Islam and Science Fiction*, 30 October 2012. http://www.islamscifi.com/interview-noura-alnoman/.

Ahram Online. 'Claims of Cure for HIV, Hepatitis C Are a "Scandal": Egypt Presidential Advisor.' *Ahram Online*, 26 February 2014. http://english.ahram.org.eg/NewsContent/1/64/95270/Egypt/Politics-/Claims-of-cure-for-HIV,-Hepatitis-C-are-a-scandal-.aspx.

———. 'Alexandria Uni cancels lecture by Egypt's ex-presidential scientific advisor Heggy.' *Ahram Online*, 24 October 2015. http://english.ahram.org.eg/NewsContent/1/64/161728/Egypt/Politics-/Alexandria-Uni-cancels-lecture-by-Egypts-expreside.aspx.

Al-Aḥmad, Barāʾ. 'Rā'id al-faḍāʾ Muḥammad Fāris: al-muṣādafah jaʿalatnī ṭayyār.' *Al-Thawrah*. 31 October 2007. http://thawra.sy/_print_veiw.asp?FileName=99074866620071030202753.

Al Arabiya. 'Rā'id faḍāʾ sūrī munshaqq: Ḥāfiẓ al-Asad hamashanī 9 sanawāt.' *Al Arabiya*, 18 March 2013. http://www.alarabiya.net/articles/2012/10/18/244443.html.

Al-Bakr, Aḥmad Ḥasan. 'Niẓām al-taṭawwuʿ fi ṣanf al-ṣawārīkh al-muwajjah (al-musayṭar ʿalayhā).' *Al-qawānīn wa-al-tashrīʿāt al-ʿIrāqīyah*, 1969. http://wiki.dorar-aliraq.net/iraqilaws/law/19758.html.

Al-Barghūthī, Samīr. 'Falakīyan. al-ʿīd ghadan.' *Al-Waṭan*, 2014. http://www.al-watan.com/viewnews.aspx?n=A9BA6017-A225-4DF5-BB7B-92EF349D06B9&d=20140727.

Al-Bayān. '"Ruwwād faḍāʾ al-Imārāt" barnāmaj mustadām li-bināʾ kawādir waṭanīyah tuʿazziz makānat al-dawlah ʿālamīyan.' *Al-Bayān*, 13 April 2017.

http://www.albayan.ae/across-the-uae/news-and-reports/2017-04-13-1.2914312.

Al-Bāz, Fārūq [see also El-Baz, Farouk]. 'Taqdīm.' In *Mādhā ba'd al-Qamar?*, edited by Ṣalāḥ Jalāl, 9–10. Al-Qāhirah: Dār al-Maʿrifah, 1975.

———. 'Dirāsat al-Arḍ min al-faḍā'.' In *Min thimār al-fikr: al-mawsim al-thaqāfī al-awwal*, 85–97. Al-Dawḥah: Kullīyat al-Tarbiyah, 1976.

———. 'Riḥalāt al-faḍā' wa-mā asdat'hu lil-insān.' *Al-'Ilm* 26 (1978): 30–35.

———. 'Taqdīm.' In *Al-ṭarīq Ilá al-Mirrīkh*, by Sa'd Shaʿbān, 9–12. Al-Kuwayt: al-Majlis al-Waṭanī lil-Thaqāfah wa-al-Funūn wa-al-Ādāb, 1997.

Al-Daybān, Nawras. '(Muḥammad Fāris) yakshif lil-Ūriyint: al-kalimah allatī azʿajat Ḥāfiẓ al-Asad.' *Orient Net*, 5 November 2014. http://orient-news.net/?page=news_show&id=82402.

Alfaisal University. 'Alfaisal faculty propose a new strategy for the advancement of Saudi national education in science, technology, engineering, and mathematics (STEM) in an article published in "The Knowledge."' *Alfaisal University*, 9 March 2014. http://news.alfaisal.edu/alfaisal-faculty-propose-a-new-strategy-for-the-advancement-of-saudi-national-education-in-science-technology-engineering-and-mathematics-stemin-an-article-published-in-the-knowledge/.

Al-Fiqī, Muṣṭafá. 'Ibrāhīm Ḥilmī 'Abd al-Raḥmān.' *Al-Maṣrī al-Yawm*, 2 April 2009. http://today.almasryalyoum.com/article2.aspx?ArticleID=205249.

———. 'Al-ʿArab wa-al-munaẓẓamāt al-duwalīyah.' *Al-Ḥayāt*, 8 March 2016. http://www.alhayat.com/m/opinion/14359691.

Al-Ghaḍawī, 'Abd Allāh. 'Al-niẓām yakhtār al-ṭayyārīn ṭā'ifīyan. wa-umniyatī qaṣf qaṣr al-Asad.' *Okaz*, 3 September 2012. http://www.okaz.com.sa/new/Issues/20120903/Con20120903529601.htm.

Al-Ḥakīm, Tawfīq. *Riḥlah ilá al-ghad*. Al-Qāhirah: Maktabat Miṣr, 1994.

Al-Ḥarīrī, Laylá. '18 'āliman 'arabīyan fawq saṭḥ al-Qamar.' *Al-Dawḥah*, no. 1 (January 1983): 84–89.

Al-Hay'ah al-ʿArabīyah lil-Taṣnīʿ. ''An al-hay'ah.' *Al-hay'ah al-ʿArabīyah lil-taṣnīʿ*, 2017. https://www.aoi.org.eg/2014-05-07-12-06-54.

Al-Hinai, Khattab G., Abdallah E. Dabbagh, Weston C. Gardner, Muhammad A. Khan, and Salih Saner. 'Shuttle Imaging Radar Views of Some Geological Features in the Arabian Peninsula.' *GeoArabia* 2, no. 2 (1997): 165–78.

Al-Ḥuṣrī, Sātiʿ. 'Ittaḥadnā fī 'aṣr al-jimāl wa-al-bighāl fa-kayf natafarraq fī 'aṣr al-dharrah wa-al-ṣawārīkh?' *Al-ʿArabī*, no. 6 (May 1959): 20–25.

Al-Jaffal, Omar. 'Iraq's defunct military industry.' Translated by Pascale el-Khoury. *Al-Monitor*, 1 July 2015. http://www.al-monitor.com/pulse/originals/2015/07/iraq-military-industry-war-islamic-state.html.

Al-Jamʿīyah al-Falakīyah al-Sūrīyah. 'Jūrj Ḥilw.' *Yawmīyāt Kawnīyah*. Accessed 11 August 2013. http://www.saaa-sy.org/diary/diary_pages/ast_george.html.

Al-Jārūshah, Yāsir. 'Qism al-falak bi-Jāmiʿat al-Malik 'Abd al-ʿAzīz yashkur al-Amīr Muqrin li-ihdā'ihi marṣad falakī.' *Al-Riyāḍ*, 28 October 2009. http://www.alriyadh.com/469826.

Al-Jazīrah. 'Al-Jiyūlūjī al-shahīr Fārūq al-Bāz.' *Maw'id fī al-mahjar*, 23 July 2001. http://www.aljazeera.net/programs/pages/0c8acb14-226d-457b-9f6f-9d0745f507ba.

Alkhaja, Alwaleed. 'Astronomy comes to Qatar.' *QScience*, 11 February 2013. http://qscienceblog.wordpress.com/2013/02/11/astronomy-comes-to-qatar/.

Al-Khalīj. 'Al-'Irāq. bidāyāt min al-'amal al-jādd wa-ikhtirāq al-faḍā'.' *Al-Khalīj*, 4 May 2016. http://www.alkhaleej.ae/supplements/page/b62408c9-d73c-4757-a469-9756e445dad8.

Al-Khalīj Ūnlāyin. 'Markaz Qaṭar li-'ulūm al-falak. inṭilāqah khalījīyah naḥw faḍā' al-ma'rifah.' *Alkhaleej Online*, 23 May 2016. http://klj.onl/2wSnzD.

Al-Khaṣāwinah, 'Awnī Muḥammad. 'Al-Markaz al-Iqlīmī li-Tadrīs wa-Tiknūlūjiyā al-Faḍā' li-Duwal Gharb Āsiyā-al-Umam al-Muttaḥidah.' *Al-Kawn* 1, no. 1 (2014): 16–19.

Al-Khedairy, Abdulaziz A. 'Saudi University Policy: King Saud Response.' *Science* 335, no. 6072 (2012): 1040.

Al-Kulliyeh. 'Two A.U.B. Professors Honored by the Lebanese Government.' *Al-Kulliyeh* 23 (January 1948): 10.

Al-Luḥaydān, Ḥamad ibn 'Abd Allāh. 'Ahammīyat inshā' wakālat al-faḍā' al-Su'ūdīyah.' *Al-Riyāḍ*, no. 15005 (24 July 2009). http://www.alriyadh.com/447092.

———. 'Ahammīyat inshā' wakālat al-faḍā' al-Su'ūdīyah (2).' *Al-Riyāḍ*, no. 15012 (31 July 2009). http://www.alriyadh.com/448744.

Al-Malki, M.B. 'Astronomy in Saudi Arabia.' In *Highlights of Astronomy*, edited by I. Appenzeller, 672–73. Dordrecht: Kluwer, 1995.

———. 'Astronomy in Saudi Arabia: The Challenges.' In *Developing Basic Space Science World-Wide: A Decade of UN/ESA Workshops*, edited by Willem Wamsteker, Rudolf Albrecht, and Hans J. Haubold, 261–66. Dordrecht: Kluwer, 2004.

Al-Maqrīzī, Taqī al-Dīn Aḥmad b. 'Alī. *Towards a Shi'i Mediterranean Empire: Fatimid Egypt and the Founding of Cairo*. Translated by Shainool Jiwa. London: I.B.Tauris, 2009.

Al-Mashat, Ali. 'The Arab Satellite Communications System.' In *AIAA 9th Communications Satellite Systems Conference*, 187–91. American Institute of Aeronautics and Astronautics, 1982.

Al-Maṣrī, Ashraf. ''Iṣām Ḥajjī: anā mas'ūl 'an inqādh al-basharīyah.' *Al-Khalīj*, 22 December 2009. http://www.alkhaleej.ae/supplements/page/f786c51b-34ff-4c89-9ea8-bd431fa5d844.

Al-Mleaky, Yasseen Mohammed. 'Density diagnostics and inhomogeneous non-isothermal plasmas.' PhD thesis, University of Glasgow, 1990.

Almleaky, Y.M., Y.E. Helali, H.M. Basurah, A.A. Malawi, A.A. Goharji, and M.N. Esmaial. 'Air Drag Effect on the Motion of an Artificial Earth Satellite.' *Earth, Moon, and Planets* 73, no. 2 (1996): 109–15.

Al-Mousli, Ahmad Tamer. 'Astronomical Site Testing with a Meteorological Satellite – The First Step for the Implementation of a Network of Robotic

Telescopes.' In *Seminars of the United Nations Programme on Space Applications: Selected Papers from Activities Held in 2000*, 73–79. New York: United Nations, 2001.

Āl Muḥammad, Anwar. 'Muqābalah maʿa rāʾid al-faḍāʾ al-ʿArabī al-Sūrī-Muḥammad Fāris.' *Qatif Astronomy Society*. Accessed 19 December 2014. http://www.qasweb.org/occasions/item.php?id=418.

Al-Muqrin, ʿĀdil ibn ʿUmar. 'Li-mādhā khasarnā al-ʿUthmān?' *Hatt Post*, 11 July 2012.

Al-Naʿīmī, Ḥamīd Mujwil [see also Al-Naimiy, Hamid M. K.]. 'Al-ʿĀbid al-ʿIrāqī yaqtaḥim al-faḍāʾ.' *Āfāq ʿArabīyah* 15, no. 1 (1989): 8–12.

———. 'Radʿ al-mashrūʿ al-Ṣuhyūnī: al-ʿIrāq wal-tiknūlūjiyā al-faḍāʾ.' *Āfāq ʿArabīyah* 15, no. 5 (1990): 18–21.

———. *Al-kawn wa-asrāruhū fī āyāt al-Qurʾān al-karīm*. Bayrūt: Arab Scientific Publishers, 2000.

Al-Naimiy, Hamid M.K. [see also al-Naʿīmī, Ḥamīd Mujwil]. 'The Iraqi National Astronomical Observatory.' *Astrophysics and Space Science* 118 (1986): 51–56.

———. 'Curriculum Vitae.' *American University of Sharjah*, May 2010. http://older.sharjah.ac.ae/English/About_UOS/AdministrativeDepartments/vcaa/Documents/Hamid%20CV%20E%20May%202010.pdf.

———. 'The role of astronomy and space sciences in Arab societies and cultures.' In *The Role of Astronomy in Society and Culture*, edited by David Valls-Gabaud and Alexander Boksenberg, 429–37. Cambridge: Cambridge University Press, 2011.

Al-Naimiy, Hamid M.K., and Hans J. Haubold. 'Preface.' *Astrophysics and Space Science* 273 (2000): vii–x.

Al-Naimiy, Hamid M.K., and Khalil Konsul. 'Astronomy and Space Sciences in Jordan.' In *Seminars of the United Nations Programme on Space Applications: Selected Papers from Activities Held in 2000*, 55–71. New York: United Nations, 2001.

———. 'Basic Space Sciences in Jordan.' In *Developing Basic Space Science World-Wide: A Decade of UN/ESA Workshops*, edited by Willem Wamsteker, Rudolf Albrecht, and Hans J. Haubold, 273–89. Dordrecht: Kluwer, 2004.

Alnajjar, Abdalla A., Ammar M. Munir, Arian L. Pregenzer, and Adriane C. Littlefield. *International Initiative to Engage Iraq's Science and Technology Community: Report on the Priorities of the Iraqi Science and Technology Community*. Albuquerque: Sandia, 2004.

Al-Najjār, Muṣṭafá. "Afwan ʿIṣām Ḥajjī anta lasta ʿantīl!' *Al-Maṣrī al-Yawm*, 14 November 2014. http://www.almasryalyoum.com/news/details/572326.

Al-Qarʿān, Hishām ʿAbd al-Jawād, Aḥmad al-Ṣuqayh, and ʿUmar ibn ʿAbd al-ʿAzīz al-Sayf. "Alá gharār mashrūʿ "Abūlū" al-Amrīkī: irsāl Suʿūdī ilá al-Mirrīkh.' *Al-Maʿrifah* 227 (February 2014): 12–17.

Al Qasim, Ibrahim H., Fatma H. Lootah, Hessa R. Almatroush, Hessa Ali, Maitha Sharif, Khalifa Al Mheiri, Shaima Al Marzooqi, Abdullah Al Shehhi, Ahmed AlShaer, and Fadya AlMaeeni. 'Nayif-1: UAE's First CubeSat

Mission.' In *SpaceOps 2016 Conference*. Daejeon: American Institute of Aeronautics and Astronautics, 2016.

Alrais, Adnan, Manfred Bester, Brett Stroozas, and Majid Alloghani. 'Emirates Mars Mission – Al Amal Overview.' In *14th International Conference on Space Operations*. Reston: American Institute of Aeronautics and Astronautics, 2016.

Al-Rajeh, Saad M. 'Sensorimotor Adaptation in Microgravity Obtained from the Saudi Astronaut During the Space Shuttle 'Discovery' 51-G Mission.' *Saudi Medical Journal* 9, no. 5 (1988): 468–75.

Al-Rashīd, Nāyif. 'Awwal rā'id faḍā' ʿArabī: 'Tshālinjir' aʿāq 5 riḥalāt mukūkīyah lil-mamlakah.' *Al-Waṭan*, 3 November 2012. http://www.alwatan.com.sa/nation/News_Detail.aspx?ArticleID=119416&CategoryID=3.

Al-Riyāḍ. 'Awāmir malakīyah bi-iʿfā' al-ʿUthmān wa-al-Rāshid wa-al-Ḥuqayl min manāṣibihim wa-taʿyīn al-ʿUmar mudīran li-Jāmiʿat al-Malik Suʿūd wa-al-Dāwud mudīran li-Jāmiʿat al-Malik Khālid.' *Al-Riyāḍ*, 2 July 2012. http://www.alriyadh.com/748442.

Al-Saʿdanī, ʿIzzat. 'Wa-kāna al-insān ẓalūman jahūlan.' *Al-Ahrām*, no. 41900 (25 August 2001). http://www.ahram.org.eg/Archive/2001/8/25/INVE1.HTM.

Al-Saʿīd, 'Muṣṭafá. 'Mamarr al-Tanmiyah wa-al-Taʿmīr wa-awlawīyāt al-tanmiyah.' *Al-Shurūq*, 5 May 2011. http://www.shorouknews.com/columns/view.aspx?cdate=05052011&id=90e4f277-9cc3-443a-9557-c5d502c4c8f5.

Alsabti, Athem. Email to John Hearnshaw. 'Visit to Iraq, including to Mt Korek Observatory,' 5 May 2004.

———. 'Astronomy in Iraq.' In *Highlights of Astronomy as presented at the IAU XXVI General Assembly, 2006*, edited by Karel van der Hucht, 655. Cambridge: Cambridge University Press, 2007.

———. 'Dr A W Alsabti.' *University College London*. Accessed 25 May 2015. http://zuserver2.star.ucl.ac.uk/~aalsabti/.

Al Saud, M.A., A.A. Malawi, H.M. Basurah, and Y.M. Al-Mleaky. 'Broad Band Filter Colours and Optical Depth of the G and L Impact Sites of Comet Shoemaker-Levy 9.' *Earth, Moon and Planets* 75 (1996): 147–55.

Al-Saud, M.A., H.M. Basurah, A.A. Malawi, and Y.M Al-Maleaky. 'Spectroscopic Observations at the Time of the SL9 Impacts.' *Journal of King Abdulaziz University: Science* 9 (1997): 53–57.

Al Saud, Sultan bin Salman bin Abdulaziz. 'Biography.' *Sultan bin Salman bin Abdulaziz*, 2013. http://www.sultanbinsalman.sa/en/Biography/Pages/deafult.aspx#tab2.

Al Saud, Sultan bin Salman bin Abdulaziz, Ahmed N. Abo-khatwa, and Tarek A. Fadaak. *One Planet: The Story of the First Arab Mission to Space*. Riyadh: Saudi Specialized Publishing, 2011.

Al-Saud, Turki Saud. '3D-Computerized Gas Emission Tomography and Particle Emission Velocimetry for Flame Measurement.' Doctor of Philosophy dissertation, Stanford University, 1997.

Al-Saud, Turki, and I-Dee Chang. 'Space Shuttle RCS Plumes: Radiation and Temperature Measurement on Mission 51-G.' In *Commercial Opportunities in Space*, edited by F. Shahrokhi, C.C. Chao, and K.E. Harwell, 462–74. Washington: American Institute of Aeronautics and Astronautics, 1988.

Al-Shamlan, Ali A. 'Introduction.' In *Atlas of the State of Kuwait from Satellite Images*, edited by Farouk El-Baz and Mohammad Al-Sarawi, 2. Safat: Kuwait Foundation for the Advancement of Sciences, 2000.

Al-Sharq al-Awsaṭ. 'Idghār Shuwayrī: Ṣuwar hubūṭ murakkibāt "Abūlū" ʿalá al-Qamar kānat al-dāfiʿ warāʾ takhaṣṣuṣī.' *Al-Sharq al-Awsaṭ*, 5 March 2006.

Al-Shaykh, ʿIṣām. 'Awwal Imārātī yazūr al-faḍāʾ khilāl ʿāmayn.' *Al-Sharq al-Awsaṭ*, 3 May 2007.

Al-Shinnāwī, Tawfīq. 'Al-ṣārūkh al-Almānī.' *Al-Thaqāfah*, no. 682 (21 January 1952): 21–22.

Alsubai, Khalid. 'Wide Angle Search for Extrasolar Planets by the Transit Method.' PhD thesis, University of St Andrews, 2008.

———. 'Astrophysics & Aerospace Workshop.' Qatar Foundation Annual Research Forum, 2010. http://www.qf-arf.org/arf2010/Workshop_on_Astrophysics_Aerospace.pdf.

Alsubai, K.A., N.R. Parley, D.M. Bramich, R.G. West, P.M. Sorensen, A. Collier Cameron, D.W. Latham, et al. 'Qatar-1b: a hot Jupiter orbiting a metal-rich K dwarf star.' *Monthly Notices of the Royal Astronomical Society* 417, no. 1 (2011): 709–16.

Alsubai, Khalid, Dimitris Mislis, Zlatan I. Tsvetanov, David W. Latham, Allyson Bieryla, Lars A. Buchhave, Gilbert A. Esquerdo, et al., 'Qatar Exoplanet Survey: Qatar-3b, Qatar-4b, and Qatar-5b.' *Astronomical Journal* 153, no. 4 (2017): 200.

Al-Subai, Khalid, and Martin Dominik. 'Conference scientific rationale.' *1st Doha International Astronomy Conference*, 2013. http://star-www.st-and.ac.uk/~md35/Doha2013/.

Al-Sudayrī, Rawḍah. 'Al-Brūfīsūr al-Naʿīmī: ʿām 2050 lan naʿīsh ʿalá al-Arḍ.' *Al-Bayān*, 4 June 2008. http://www.albayan.ae/our-homes/1212501142739-2008-06-04-1.644619.

Altonn, Helen. 'UH astronomers observe eclipse in Libyan desert.' *Star Bulletin* 11, no. 107 (2006). http://archives.starbulletin.com/2006/04/17/news/story02.html.

Al-ʿUlabī, Ibrāhīm. 'Min al-turāth ilá al-samāʾ. huwwāt al-falak al-Khalījīyūn ẓāhirah qadīmah tatajaddad.' *Alkhaleej Online*, 3 June 2016. http://klj.onl/Zql4sR.

Al-ʿUmarī, ʿAbd al-ʿAzīz. 'Kayfa yuṭlaq ʿalaykum musammá "falakīyīn" wa-antum lā taḥmilūn shahādāt mutakhaṣṣiṣah?' *Al-Yawm*, 27 April 2014. http://www.alyaum.com/article/3135671.

Aly, M.K. 'Khartoum Expeditions for Total Solar Eclipse of February 25th, 1952.' *Observatory* 72, no. 867 (1952): 63–72.

———. 'Preliminary Note on Measures of Coronal Emission Lines Observed at the Total Solar Eclipse, February 25, 1952, by B. Lyot and M.K. Aly.' *Astrophysical Journal* 122 (1955): 438–44.

Aly, M.K., J.W. Evans, and F.Q. Orrall. *A Spectrographic Atlas of the Solar Corona from λ3300 to λ6900 Å*. Sunspot: Sacramento Peak Observatory and Helwan Observatory, 1962.

———. 'A Photometric Study of the Continuum and Seventeen Emission Lines in the Inner Solar Corona.' *Astrophysical Journal* 136 (1962): 956–74.

———. 'Emission-Line Ratios in the Solar Corona.' *Astrophysical Journal* 137 (1963): 1313–16.

Aly, M. Khairy, and A.A. Galal. 'Choice of a Site for the Astronomical Station of King Abdulaziz University at Jeddah.' *Bulletin of the Faculty of Science, King Abdul Aziz University* 6 (1982): 177–95.

Aman, Ayah. 'Egyptian satellite will monitor water security, Renaissance Dam.' *Al-Monitor*, 8 May 2014. http://www.al-monitor.com/pulse/originals/2014/05/egypt-satellite-launch-space-program-renaissance-dam.html.

American University of Beirut. 'The American University of Beirut Awards Six Honorary Doctorates in 2003.' *AUB*, 28 May 2003. http://www.aub.edu.lb/news/Pages/29696.aspx.

———. 'AUB mourns Angela Jurdak Khoury.' *AUB*, 30 May 2011. http://www.aub.edu.lb/news/Pages/angela-khoury.aspx.

———. 'Lee Observatory.' *AUB*. Accessed 9 August 2013. http://www.aub.edu.lb/tour/nojava/b12.html.

Amīn, Ḥussām Muḥammad. 'Al-ʿĀbid.' *Shabakat al-Baṣrah*, 28 April 2011. http://www.albasrah.net/ar_articles_2011/0411/hosam_280411.htm.

Arab News. 'Planet-finder Qatar over the moon.' *Arab News*, 15 December 2010. http://www.arabnews.com/node/363144.

Arab Union for Astronomy and Space Sciences. *Al-mu'tamar al-falakī al-Islāmī al-rābiʿ: kutayyib mulakhkhaṣāt*. Amman: AUASS, 2006.

Asaad, A., B. Baghous, M. Fahim, H. Jelinkova, A. Novotny, K. Hamal, S. Tatavian, et al. 'A Joint Interkosmos–SAO–HIAG Laser Ranging Project.' In *The Use of Artificial Satellites for Geodesy and Geodynamics*, edited by George Veis and Evangelos Livieratos, 2: 94–106. Athens: National Technical University of Athens, 1979.

Asaad, Adly Salama, and Joseph Sidky Mikhail. 'Colour Contrasts of Lunar Grounds.' *Moon* 11, no. 3–4 (1974): 273–99.

———. 'Lunar Work Carried Out at Kottamia Observatory.' *Moon and the Planets* 20, no. 2 (1979): 153–55.

Associated Press. 'Star Gazing in Saudi.' *Daytona Beach Morning Journal*. 14 September 1967.

Association of Space Explorers. 'I Congress.' *Association of Space Explorers*, 1985. http://www.space-explorers.org/congress/congress1.html.

———. 'V Congress General Statement,' *Association of Space Explorers*, 1989. http://www.space-explorers.org/congress/statements/statement5.html.

———. 'Statement on Universal Space Rescue.' *Association of Space Explorers*, 1989. http://www.space-explorers.org/congress/statements/usr.html.

ʿAṭīyah, Ḥusām. '"Al-Dustūr" tuḥāwir thānī rāʾid faḍāʾ ʿArabī wa-awwal Sūrī yaṣʿud ilá al-faḍāʾ.' *Al-Dustūr*, 10 August 2009.

Atkinson, R. d'E. 'Greenwich Eclipse Expedition.' *Sky and Telescope* 12 (November 1952): 9–10.

Austen, Kat. 'To Infinity and beyond.' *New Scientist* 209, no. 2803 (12 March 2011): 53.

Baars, Jacob W.M. *International Radio Telescope Projects: A Life among Its Designers, Builders and Users*. CreateSpace, 2013.

Bachelet, E., I.-G. Shin, C. Han, P. Fouqué, A. Gould, J.W. Menzies, J.-P. Beaulieu, et al., 'MOA 2010-BLG-477Lb: Constraining the Mass of a Microlensing Planet from Microlensing Parallax, Orbital Motion, and Detection of Blended Light.' *Astrophysical Journal* 754, no. 1 (2012): 73.

Badawī, Rāniyā. ''Iṣām Ḥajjī fī ḥiwār lil-"Maṣrī al-yawm": Maṣr taʿīsh fī mashhad "ʿashwāʾī" yughaddī al-irhāb.' *Al-Maṣrī al-Yawm*, 14 February 2015. http://www.almasryalyoum.com/news/details/657137.

Bass, Thomas A. *Reinventing the Future: Conversations with the World's Leading Scientists*. Reading: Addison-Wesley, 1994.

Basurah, Hassan Mohammed. 'Spectropolarimetric studies of the sun, the moon and the daytime sky.' PhD thesis, University of Glasgow, 1991.

Bates, Brainerd S. 'Space-Age Immigrant.' *Aramco World* 27, no. 6 (1976): 18–25.

Becker, Uwe. 'Saudi University Policy: Meaningful Cooperation. *Science* 335, no. 6072 (2012): 1041.

Behar, Richard. 'There He Goes Again: Egypt's Morsi Stuns U.S. Senators In Meeting With "Jews-Control-Media" Slur.' *Forbes*, 25 January 2013. http://www.forbes.com/sites/richardbehar/2013/01/25/there-he-goes-again-egypts-morsi-stuns-u-s-senators-in-meeting-with-jews-control-media-slur/.

Bennett, W. Tapley. 'USUN Daily Classified Summary No. 125.' *WikiLeaks*, 3 July 1974. https://wikileaks.org/plusd/cables/1974USUNN02330_b.html.

Berloty, P. 'L'Observatoire de Ksara (Liban).' *Ciel et Terre* 33 (1912): 103–09.

Bhattacharjee, Yudhijit. 'Saudi Universities Offer Cash In Exchange for Academic Prestige.' *Science* 334, no. 6061 (2011): 1344–45.

Bingham, Millicent Todd. 'Solar Eclipse Photography. A Forerunner of the Motion Picture Camera.' *Popular Astronomy* 31 (1923): 631–34.

Blamont, Jacques. 'Les premières expériences d'aéronomie en France.' In *L'Essor des recherches spatiales en France*, edited by Brigitte Schürmann, 31–41. Noordwijk: European Space Agency, 2001.

Bliss, Daniel. *The Reminiscences of Daniel Bliss*. New York: Revell, 1920.

Bliss, Howard S. 'The Modern Missionary.' *The Atlantic Monthly* (May 1920): 664–75.

Bohannon, John. 'A challenge to pseudoscience.' *Science* 345, no. 6192 (2014): 16.

Bornmann, Lutz, and Johann Bauer. 'Which of the World's Institutions Employ the Most Highly Cited Researchers? An Analysis of the Data From Highlycited.com.' *Journal of the Association for Information Science and Technology* 66, no. 10 (2015): 2146–48.

Bramich, D.M., K.A. Alsubai, A. Arellano Ferro, N.R. Parley, A. Collier Cameron, K. Horne, D. Pollacco, and R.G. West. 'RR Lyrae Stars in the GCVS Observed by the Qatar Exoplanet Survey.' *Information Bulletin on Variable Stars* 6106 (2014).

Brandenstein, Daniel C., and Carol C. Butler. 'Oral History Transcript.' *NASA Johnson Space Center Oral History Project*, 19 January 1999. http://www.jsc.nasa.gov/history/oral_histories/BrandensteinDC/DCB_1-19-99.pdf.

Bray, R.A. 'A Remarkable Flight of Birds.' *Nature* 52, no. 1348 (1895): 415.

Broad, William J. 'Shuttle Lofts Satellite for Arab Lands and P.L.O.' *New York Times*, 19 June 1985.

Bryan, Marta L., Khalid A. Alsubai, David W. Latham, Neil R. Parley, Andrew Collier Cameron, Samuel N. Quinn, Joshua A. Carter, et al., 'Qatar-2: A K Dwarf Orbited by a Transiting Hot Jupiter and a More Massive Companion in an Outer Orbit.' *Astrophysical Journal* 750, no. 1 (2012): 84.

Bū al-Jidarī, Yāsmīn. 'Al-falakī al-Jazā'irī Niḍāl Qassūm lil-Naṣr: Wazārat al-Shu'ūn al-Dīnīyah irtakabat akhṭā' jasīmah fī raṣd al-ahillah wa-lā yaṣaḥḥ itbā' al-Su'ūdīyah.' *Al-Naṣr*, 6 March 2015. http://www.annasronline.com/index.php/2014-08-17-13-22-10/vip/4605-2015-03-06-22-09-00.

Cable News Network. ''Iṣām Ḥajjī ya'tadhir 'an iddi'ā' ḥadhf sīratihi min kitāb madrasī Miṣrī: i'itamadtu 'alá maṣādir i'tadtu al-thiqah bihā.' *CNN Arabic*, 23 March 2016. http://arabic.cnn.com/middleeast/2016/03/23/egypt-essam-heggy-education-ministry.

———. 'Al-Jazā'ir wa-al-Imārāt tuwassi'ān ta'āwunahumā fī majāl al-faḍā'.' *CNN Arabic*, 25 April 2017. https://arabic.cnn.com/world/2017/04/25/algeria-uae-space-science.

Cairo Post. 'Communication with EgyptSat 2 Satellite Lost: Russian Newspaper.' *Cairo Post*, 24 April 2015. http://thecairopost.youm7.com/news/147495/news/communication-with-egyptsat-2-satellite-lost-russian-newspaper.

Campbell, W.W., and C.D. Perrine. 'The Lick Observatory–Crocker Eclipse Expedition to Spain.' *Publications of the Astronomical Society of the Pacific* 18 (1906): 13–36.

Carpenter, Frederic W. 'An Astronomical Determination of the Heights of Birds during Nocturnal Migration.' *Auk* 23, no. 2 (1906): 210–17.

Centre de Recherche en Astronomie, Astrophysique et Géophysique. 'Jean Coulomb (1904–1999).' *La lettre du CRAAG*, no. 11 (2002): 4.

———. 'Charles Trépied (1845–1907).' *La lettre du CRAAG*, no. 12 (2003): 4.

———. 'Commémoration du 30ème anniversaire de la création du CRAAG.' *La lettre du CRAAG*, no. 50 (2010): 6.

———. 'Jean Lagrula (1906–1988), Ex Directeur IMPGA (1961–1968).' *CRAAG Infos*, no. 52 (2011): 3.

Chad, Sheldon. 'The Forgotten Apogee of Lebanese Rocketry.' *Saudi Aramco World* 64, no. 3 (2013): 18–23.

Chamberlain, J.M., and Owen Gingerich. 'Eclipse in Ceylon.' *Sky and Telescope* 14 (September 1955): 452–66.

Choudry, Ash. 'Interview with Prof Farouk El-Baz.' *Nottingham Science Blog*, 27 August 2012. http://nottinghamscience.blogspot.com/2012/08/interview-with-prof-farouk-el-baz.html.

Christie, John, and Patricia Moody. 'Arabsat – the Impact.' *Aramco World* 36, no. 2 (1985): 18–23.

Clark, Arthur [see also Klārk, Arthar]. 'Arabsat: A Dream Fulfilled.' *Aramco World* 36, no. 2 (1985): 13.

Clément, G., T. Viéville, F. Lestienne, and A. Berthoz. 'Modifications of Gain Asymmetry and Beating Field of Vertical Optokinetic Nystagmus in Microgravity.' *Neuroscience Letters* 63, no. 3 (1986): 271–74.

Clément, G., and A. Berthoz. 'Cross-Coupling between Horizontal and Vertical Eye Movements during Optokinetic Nystagmus and Optokinetic After-nystagmus Elicited in Microgravity.' *Acta Otolaryngologica* 109, no. 3–4 (1990): 179–87.

Clément, G., and F. Lestienne. 'Adaptive modifications of postural attitude in conditions of weightlessness.' *Experimental Brain Research* 72, no. 2 (1988): 381–89.

Conference of Heads of State or Government of the Non-Aligned Countries. *Programme for Peace and International Co-Operation: Declaration as Adopted by the Conference*, 10 October 1964. http://cns.miis.edu/nam/documents/Official_Document/2nd_Summit_FD_Cairo_Declaration_1964.pdf.

Cook, Jean Marie. 'AUB announces this year's honorary doctorate recipients: Mary Robinson, Marcel Khalife, Owen Gingerich, Mostafa El-Sayed, and Anthony Shadid.' *American University of Beirut*, 27 May 2011. http://www.aub.edu.lb/news/Pages/hd-2011-announcement.aspx.

COPUOS. 'Report of the Committee on the Peaceful Uses of Outer Space.' 1968. Twenty-third session, Agenda item 24, A/7285. United Nations General Assembly Official Records.

———. 'Verbatim Record of the One Hundred and Thirty-First Meeting,' July 1, 1974. A/AC.105/PV. 131. United Nations General Assembly Official Records. http://www.unoosa.org/pdf/transcripts/copuos/AC105_PV131E.pdf.

———. *Report on the United Nations/European Space Agency Workshop on Basic Space Science for Developing Countries, Organized in Cooperation with the Government of India and Hosted by the India Space Research Organization.* 17 September 1991. A/AC.105/489. United Nations General Assembly Official Records.

Corbett, Ian F., ed. *Proceedings of the Twenty Seventh General Assembly, Rio de Janeiro, Brazil, 2009*, 355–419. Transactions of the International Astronomical Union XXVIIB. Cambridge: Cambridge University Press, 2010.

Cordall, Simon Speakman. 'Star Wars' abandoned Tunisian locations – in pictures.' *The Guardian*, 3 December 2015, sec. Film. https://www.theguardian.com/film/2015/dec/03/star-wars-abandoned-tunisian-locations-in-pictures.

Crawford, David L. 'GNAT: Global Network of Automatic Telescopes.' In *Automated Telescopes for Photometry and Imaging*, edited by Saul J. Adelman, Robert J. Dukes, and Carol J. Adelman, 111–27. San Francisco: Astronomical Society of the Pacific, 1992.

———. 'GNAT – A Global Network of Automatic Telescopes.' In *Stellar Photometry: Current Techniques and Future Developments*, edited by C.J. Butler and I. Elliott, 244–49. Cambridge: Cambridge University Press, 1993.

Creighton, John O., and Jennifer Ross-Nazzal. 'Oral History Transcript.' *NASA Johnson Space Center Oral History Project*, 3 May 2004. http://www.jsc.nasa.gov/history/oral_histories/CreightonJO/CreightonJO_5-3-04.pdf.

Croucher, Martin. 'Work to start on RAK observatory this year.' *The National*, 14 May 2014. http://www.thenational.ae/uae/science/work-to-start-on-rak-observatory-this-year.

Culbertson, Shelly, Michael G. Mattock, Bruce R. Nardulli, Abdulrazaq Al-Kuwari, Gary Cecchine, Margaret C. Harrell, John A. Friel, and Richard E. Darilek. *Launching the Qatar National Research Fund*. Santa Monica: RAND Corporation, 2012.

Dabbagh, Abdallah E., Khattab G. Al-Hinai, and M. Asif Khan. 'Detection of Sand-Covered Geologic Features in the Arabian Peninsula Using SIR-C/X-SAR Data.' *Remote Sensing of Environment* 59, no. 2 (1997): 375–82.

Damiens, Jacques. 'La météorologie française au Levant (1921–1929).' *La Météorologie* (April 1995): 28–34.

Damon, Arwa, and Tim Lister. 'Tunisian "Star Wars" town now features in ISIS battle.' *CNN*, 24 March 2015. http://www.cnn.com/2015/03/23/africa/tunisia-fallout/index.html.

De Jager, C., and A. Jappel, eds. *Proceedings of the Fourteenth General Assembly, Brighton 1970*, 3–28. Dordrecht: Reidel, 1971.

De Vaucouleurs, G., J. Blunck, M. Davies, A. Dollfus, I.K. Koval, G.P. Kuiper, H. Masursky, et al., 'The New Martian Nomenclature of the International Astronomical Union.' *Icarus* 26, no. 1 (1975): 85–98.

Driessen, Henk. 'Mediterranean Port Cities: Cosmopolitanism Reconsidered.' *History and Anthropology* 16, no. 1 (2005): 129–41.

Durra, Yasar, and John Christie. 'Arabsat – the Launch.' *Aramco World* 36, no. 2 (1985): 10–17.

Egyptian Streets. 'Egypt Plans to Build First City on Mars.' *Egyptian Streets*, 1 April 2017. https://egyptianstreets.com/2017/04/01/egypt-plans-to-build-first-city-on-mars/.

El-Awady, Nadia. 'Qatar discovers three new planets.' *Nature Middle East*, 7 July 2016. http://www.natureasia.com/en/nmiddleeast/article/10.1038/nmiddleeast.2016.105.

El-Baz, Farouk [see also al-Bāz, Fārūq]. *Say It in Arabic (Egyptian Dialect).* New York: Dover, 1968.

———. 'Discovery of Two Lunar Features.' In *Apollo 16 Preliminary Science Report*, 33–38. Washington: NASA, 1972.

———. 'Al-Khwarizmi: A New-Found Basin on the Lunar Far Side.' *Science* 180, no. 4091 (1973): 1173–76.

———. 'Expanding desert creates grim beauty but also threatens crucial cropland.' *Smithsonian* 8, no. 3 (1977): 34–41.

———. 'Earth Orbital Photography by the Large Format Camera.' In *Twelfth International Symposium on Remote Sensing of Environment, April 20–26, 1978, Manila, Philippines: Summaries*, 71. Ann Arbor: Environmental Research Institute of Michigan, 1978.

———. 'Narrative of the Journey.' *Geographical Journal* 146, no. 1 (1980): 51–59.

———. 'Journey into Northern China: Impressions of a Scientist.' *Explorers Journal* 58, no. 2 (1980): 60–65.

———. 'Interdisciplinary Research at the Boston University Center for Remote Sensing.' In *Electronic Imaging East 89: Advanced Printing of Paper Summaries*, 2:1131–37. MG Expositions Group, 1989.

———. 'Do people make deserts?' *New Scientist* 128, no. 1738 (1990): 41–44.

———. 'Zhao's Impact on the Understanding of Chinese Geography in the U.S.A.' In *Selected Works of Zhao Sungqiao*, 468–70. Beijing, 1998.

———. *The Arab World and Space Research: Where Do We Stand?* Abu Dhabi: Emirates Center for Strategic Studies and Research, 1998.

———. 'Training *Apollo* astronauts in lunar orbital observations and photography.' *Geological Society of America Special Papers* 483 (2011): 49–65.

———. 'Dr. Farouk El-Baz, Director.' *Boston University*. Accessed 3 December 2013. http://www.bu.edu/remotesensing/faculty/el-baz/.

———. 'Biography.' *Dr. Farouk El-Baz*. Accessed 1 February 2014. http://faroukelbaz.com/index.php?option=com_content&view=article&id=113&Itemid=2.

———. 'Development Corridor.' *Dr. Farouk El-Baz*. Accessed 3 February 2014. http://faroukelbaz.com/index.php?option=com_content&view=article&id=22&Itemid=59.

———. '1001 Wells for Darfur.' *Boston University*. Accessed 4 February 2014. http://www.bu.edu/remotesensing/research/dafur/.

———. 'GSA Student Award named for Professor Farouk El-Baz.' *Dr. Farouk El-Baz*. Accessed 18 February 2015. http://faroukelbaz.com/index.php?option=com_content&view=article&id=5&Itemid=57.

———. 'Dr. Farouk El Baz: Public Figure.' *Facebook*. Accessed 15 March 2017. https://www.facebook.com/ELBAZ.F/.

———. 'A Street in Ras Al-Khaima, UAE.' *Dr. Farouk El-Baz*. Accessed 5 November 2016. http://faroukelbaz.com/.

El-Baz, Farouk, A.M. Worden, and V.D. Brand. 'Astronaut observations from lunar orbit and their geologic significance.' In *Proceedings of the Third Lunar Scientific Conference*, 3:85–104. Cambridge: MIT Press, 1972.

El-Baz, Farouk, and Mohamed A. Gheith. 'Memorial of Riad Abdel-Megid Higazy.' *American Minerologist* 55, no. 3–4 (1970): 569–73.

El-Baz, Farouk, and D.A. Mitchell. 'Remote Sensing as a Tool for Development.' Preprint of an invited paper to be delivered at the First Islamic Solidarity Conference in Science and Technology, University of Riyadh, 1975.

El-Baz, Farouk, and Ted A. Maxwell. *Desert Landforms of Southwest Egypt: A Basis for Comparison with Mars*. Washington: NASA, 1982.

El-Baz, Farouk, and S.A. Roosa. 'Significant results from Apollo 14 lunar orbital photography.' In *Proceedings of the Third Lunar Scientific Conference*, 1:63–83. Cambridge: MIT Press, 1972.

El-Baz, Farouk, and Rebecca Wright. 'Farouk El-Baz.' *NASA Johnson Space Center Oral History Project*, 2 November 2009. http://www.jsc.nasa.gov/history/oral_histories/El-BazF/El-BazF_11-2-09.htm.

El-Baz, Pat O'Leary. 'Pat O'Leary El-Baz Writes.' *Swampscott (Massachusetts) High School Class of 1953 Biographical Notes*, 23 May 2003. http://www.cowbeech.force9.co.uk/biog.htm.

El-Behary, Hend. 'Omar Samra to become first Egyptian in space.' *Egypt Independent*, 11 December 2013. http://www.egyptindependent.com/news/omar-samra-become-first-egyptian-space.

El-Faer, M.Z., Mohammad Farhat Ali, Hamza K. Asar, and Sultan Ibn Salman Al-Saud. 'Phase Separation in Microgravity—Evaluation of ARABSAT Phase Separation Experiment Results.' *Arabian Journal for Science and Engineering* 13, no. 3 (1988): 355–67.

El Kony, Mohamed Awad, 'Letter.' 28 October 1964. http://cns.miis.edu/nam/documents/Official_Document/2nd_Summit_FD_Cairo_Declaration_1964.pdf.

Emery, Chris. 'Choueiri elected President of the Lebanese Academy of Sciences.' *Princeton University*, 9 October 2008. http://www.princeton.edu/engineering/news/archive/?id=135.

Es'hailSat. 'Timeline.' *EshailSat*, 2016. https://www.eshailsat.qa/en/Dynamic-Pages/aboutus/#tab-5.

Fabian, John M., and Jennifer Ross-Nazzal. 'Oral History Transcript.' *NASA Johnson Space Center Oral History Project*, 10 February 2006. http://www.jsc.nasa.gov/history/oral_histories/FabianJM/FabianJM_2-10-06.pdf.

Fahmy, Dalia. 'Pioneers in a former wasteland.' *Financial Times*, 19 February 2005. http://www.ft.com/cms/s/0/934924bc-821c-11d9-9e19-00000e2511c8.html.

Fandīk, Kurnīlīyūs. *Kitāb al-mir'āh al-waḍīyah fī al-kurah al-Arḍīyah*. Bayrūt, 1852.

Fares, M.A., and G. Edelby. 'TM-3: A Joint Soviet–Syrian Space Mission.' In *History of Rocketry and Astronautics: Proceedings of the Thirty-First History*

*Symposium of the International Academy of Astronautics*, edited by Donald C. Elder and George S. James, 225–29. San Diego: Univelt, 2005.

Farooqui, Suhail Z., Shaukat N. Goderya, and K.R. Fawz-ul-Haq. 'Establishment of an Observatory Housing a 1.5-Metre Telescope – The Case for Pakistan to Join NORT.' In *Seminars of the United Nations Programme on Space Applications: Selected Papers from Activities Held in 2001*, 77–86. New York: United Nations, 2002.

Fawzī, Mufīd. 'Al-qamar al-ṣināʿi al-ʿArabī bayn al-maḥādhīr al-siyāsīyah wa-al-makhāwif al-nafsīyah!' *Al-Dawḥah*, no. 8 (August 1986): 68–71.

Fesenkov, V.G. 'The Scientific Work of the Helwan Observatory.' *Soviet Astronomy* 2 (1958): 256–59.

———. 'Expedition of the Academy of Sciences of the USSR to Aswan, Egypt, for Observations of the Zodiacal Light and of the Optical Properties of the Atmosphere.' *Soviet Astronomy* 2 (1958): 276–85.

Filasṭīn al-Yawm. 'Thalāthat "tiliskūbāt" falakīyah mutaṭawwirah taṣil Ghazzah.' *Filasṭīn al-Yawm*, 18 March 2013. https://paltoday.ps/ar/post/162955/.

Flanagan, Ben. 'Jordan theme park treks ahead.' *The National*, 1 August 2011. http://www.thenational.ae/business/travel-tourism/jordan-theme-park-treks-ahead.

Garthwaite, Rosie. 'From astronaut to refugee: how the Syrian spaceman fell to Earth.' *The Guardian*, 1 March 2016. http://www.theguardian.com/world/2016/mar/01/from-astronaut-to-refugee-how-the-syrian-spaceman-fell-to-earth.

Geological Society of America. 'Farouk El-Baz Award for Desert Research.' *Geological Society of America*, 18 November 2013. http://www.geosociety.org/qgg/FEBA.html.

George, D., A.P. Hartsuijker, and M.A. Kaftan-Kassim. 'High Resolution Radio Interferometric Observations of the Planetary Nebulae NGC 40, NGC 6543, and NGC 6720.' *Astronomy and Astrophysics* 35 (1974): 219–24.

Gerbaldi, Michèle. 'International Schools for Young Astronomers (ISYA): A Programme of the International Astronomical Union.' In *Astronomy for the Developing World*, edited by J.B. Hearnshaw and P. Martinez, 221–28. Cambridge: Cambridge University Press, 2007.

Ghazal, Rym. 'The first Syrian in space.' *The National*, 9 April 2015. http://www.thenational.ae/arts-lifestyle/the-first-syrian-in-space.

Gillon, Michaël, Amaury H.M.J. Triaud, Brice-Olivier Demory, Emmanuël Jehin, Eric Agol, Katherine M. Deck, Susan M. Lederer, et al., 'Seven temperate terrestrial planets around the nearby ultracool dwarf star TRAPPIST-1.' *Nature* 542, no. 7642 (2017): 456–60.

Gingerich, Owen. 'Eclipse Experiences.' *Sky and Telescope* 10 (October 1951): 287–88.

———. 'Eclipse Saros Series.' *Sky and Telescope* 10 (October 1951): 288.

———. 'American University Observatory.' *Sky and Telescope* 16, no. 5 (1957): 212–23.

———. 'The Naming of Uranus and Neptune.' *Astronomical Society of the Pacific Leaflets* 8, no. 352 (1958).

———. 'A Model of Jupiter's Satellite Orbits.' *Sky and Telescope* 18, no. 7 (1959): 376–79.

———. 'PhD language requirement.' *Physics Today* 30, no. 11 (1977): 9–11.

———. 'Islamic Astronomy.' *Scientific American* 254, no. 4 (1986): 74–83.

———. 'Zoomorphic Astrolabes and the Introduction of Arabic Star Names into Europe.' *Annals of the New York Academy of Sciences* 500, no. 1 (1987): 89–104.

———. 'Five Centuries of Astronomical Textbooks and Their Role in Teaching.' In *The Teaching of Astronomy*, edited by J.M. Pasachoff and J.R. Percy, 189–95. Cambridge: Cambridge University Press, 1990.

———. 'Through Rugged Ways to Galaxies.' *Journal for the History of Astronomy* 21 (1990): 77–88.

———. 'Vita.' *Uniwersytet Zielonogórski*, 2004. http://www.bu.uz.zgora.pl/bu/gingerich/pdf/vita.pdf.

———. 'Campus Constellations.' *MainGate* 10, no. 4 (2011): 1–6.

Gingerich, Owen, and David DeVorkin. 'Oral History Transcript.' *American Institute of Physics*, 2005. http://www.aip.org/history/ohilist/31886.html.

Glanz, James. 'Two Spacecraft Track the Solar Wind to Its Source.' *Science* 278, no. 5337 (1997): 387–88.

Guessoum, Nidhal [see also Qassūm, Niḍāl]. 'Algeria.' *Mercury* 25 (April 1996): 5.

———. *Islam's Quantum Question: Reconciling Muslim Tradition and Modern Science*. London: I.B.Tauris, 2011.

———. 'Time for an Arab astronomy renaissance.' *Nature* 498, no. 7453 (2013): 161–64.

———. 'Arab Space Programs Slowly Coming of Age.' *Muslim Science*, 1 October 2014. http://muslim-science.com/arab-space-programs-slowly-coming-age/.

———. 'The future of Arab astronomy.' *Nature Middle East*, 22 October 2014. http://www.natureasia.com/en/nmiddleeast/article/10.1038/nmiddleeast.2014.258.

Guessoum, Nidhal, and Kiram Meziane. 'Visibility of the thin lunar crescent: the sociology of an astronomical problem (A case study).' *Journal of Astronomical History and Heritage* 4, no. 1 (2001): 1–14.

Gulf News. 'UAE unveils details of UAE Mars mission.' *Gulf News*, 6 May 2015. http://gulfnews.com/news/uae/government/uae-unveils-details-of-uae-mars-mission-1.1505710.

———. 'UAE to build first city on Mars by 2117.' *Gulf News*, 14 February 2017. http://m.gulfnews.com/news/uae/general/uae-to-build-first-city-on-mars-by-2117-1.1978549.

Habbal, S.R., J. Johnson, P. Nisenson, R. Woo, S. Fineschi, R. Esser, C.H. Wood, et al., 'Temperature, density and magnetic field structure of the corona during the total eclipse of 1999 August 11.' *Bulletin of the American Astronomical Society* 32 (2000): 817.

Habbal, S.R., R. Woo, S. Fineschi, R. O'Neal, J. Kohl, G. Noci, and C. Korendyke. 'Origins of the Slow and the Ubiquitous Fast Solar Wind.' *Astrophysical Journal Letters* 489, no. 1 (1997): L103–6.

Habbal, Shadia Rifai, Huw Morgan, Judd Johnson, Martina Belz Arndt, Adrian Daw, Sarah Jaeggli, Jeff Kuhn, and Don Mickey. 'Localized Enhancements of $Fe^{+10}$ Density in the Corona as Observed in Fe XI 789.2 Nm during the 2006 March 29 Total Solar Eclipse.' *Astrophysical Journal* 663, no. 1 (2007): 598–609.

Habgood, F. 'Report on a Visit to CAIRO, January 1982.' Report no. 1982/1. Nottingham: Institute of Geological Sciences, 1982.

Ḥāfiẓ, Muḥammad. 'Muʾtamar al-falak yastaʿriḍ injāzāt Qaṭar.' *Al-Rāyah*, 11 February 2013. http://www.raya.com/news/pages/7d77cb8e-c693-4abd-9321-d9e196c1bcf7.

Hajjar, Roger. 'Astronomy in Lebanon: Past, Present and Future Prospects.' In *Seminars of the United Nations Programme on Space Applications: Selected Papers from Activities Held in 2000*, 12:49–54. New York: United Nations, 2001.

Hajjar, Roger, François Querci, Monique Querci, Hamid Al-Naimiy, and Khalil Konsul. 'The Network of Oriental Robotic Telescopes (NORT): An Arab Project.' In *Seminars of the United Nations Programme on Space Applications: Selected Papers from Activities Held in 2000*, 43–48. 12. New York: United Nations, 2001.

Hamid, S.E. *First-Order Planetary Theory: Perturbations in Rectangular Coordinates Hansen's Variables, Longitude, and Distance.* SAO Special Report 285. Cambridge: Smithsonian Astrophysical Observatory, 1968.

Hamid, Salah E., and Fred L. Whipple. 'Common Origin between the Quadrantids and the δ Aquarids Streams.' *Astronomical Journal* 68 (1963): 537.

———. Hamad, Salah. Letter to Chairman, Department of Economics, Harvard University, 6 May 1964. RU007431, Fred Lawrence Whipple Papers, circa 1927–1983, Box 6, Folder: Hamid, Salah E. Smithsonian Institution Archives.

———. 'Tabular Planetary Positions from 500 B.C. to A.D. 2000.' *Astronomical Journal Supplement* 73 (1968): 16–17.

Hamid, Salah, C.M. Huffer, and Zdeněk Kopal. 'Computation of elements of eclipsing binaries by automatic methods.' *Astronomical Journal* 55, no. 1186 (1950): 169.

Hanania, Ray. 'Christa McAuliffe, an American Arab in shuttle disaster.' *Arab Daily News*, 28 January 2016. http://thearabdailynews.com/2016/01/28/christa-mcauliffe-american-arab-in-shuttle-disaster/.

Hare, Julie. 'Saudis Recruit to Hijack Rankings.' *The Australian*, 21 March 2012. http://www.theaustralian.com.au/higher-education/saudis-recruit-to-hijack-rankings/story-e6frgcjx-1226305563023.

Haubold, Hans J. 'UN/ESA workshops on basic space science: an initiative in the world-wide development of astronomy.' *Journal of Astronomical History and Heritage* 1, no. 2 (1998): 105–21.

Hearnshaw, John. 'Commission 46: Program Group for the World-Wide Development of Astronomy.' In *Reports on Astronomy: Proceedings of the International Astronomical Union*, edited by Oddbjørn Engvold, 387–90. Cambridge: Cambridge University Press, 2005.

Heggy, Essam. 'Etude et modélisation des performances des radars sondeurs basse fréquence pour la recherche de l'eau dans le sous-sol de Mars.' Thèse de doctorat, Université de Paris VI, 2002.

———. 'Essam Heggy.' *Facebook*. Accessed 15 March 2017. https://www.facebook.com/essam.heggy.

Helou, George. 'SIRTF Profiles: George Helou.' *Spitzer Space Telescope*. Accessed 10 August 2013. http://legacy.spitzer.caltech.edu/features/p_georgehelou.shtml.

———. 'Dr. George Helou.' *Académie des Sciences du Liban*. Accessed 9 April 2014. http://asduliban.org/Helou.html.

Herbig, George. 'Alfred Harrison Joy.' *Quarterly Journal of the Royal Astronomical Society* 15 (1974): 526–31.

Hiddleston, Sarah. 'Nearby star hosts seven planets that may sustain life.' *Nature Middle East*, 22 February 2017. http://www.natureasia.com/en/nmiddleeast/article/10.1038/nmiddleeast.2017.37.

Hijris, Muḥammad. 'Rā'id al-faḍā' "al-Fāris" ra'īsan li-ḥukūmat Sūriyā al-ḥurrah.' *Al-Yawm*, 25 February 2013. http://www.alyaum.com/article/3073095.

Hill, Frank, George Fischer, Suzanne Forgach, Jennifer Grier, John W. Leibacher, Harrison P. Jones, Patricia B. Jones, et al., 'The Global Oscillation Network Group Site Survey.' *Solar Physics* 152, no. 2 (1994): 351–79.

Hudson, Charles J. 'David Todd, 1855–1939: An Appreciation.' *Popular Astronomy* 47 (1939): 472–76.

Ḥusayn, Ṣaddām. 'Qānūn al-ta'dīl al-rābi' li-qanūn majlis al-baḥth al-'ilmī raqm 172/ li-sanat 1980.' *Al-qawānīn wa-al-tashrī'āt al-'Irāqīyah*, 1988. http://wiki.dorar-aliraq.net/iraqilaws/law/7360.html.

Ḥusnī, Iqbāl. 'Fī dā'irat al-ḍaw'. D. Ṣafā' al-Bāz.' *Al-Ahrām* 125, no. 41776 (23 Apri, 2001). http://www.ahram.org.eg/archive/2001/4/23/WOMN1.HTM.

Hussey, Ethel Fountain. 'The Recent Eclipse in Egypt.' *Popular Astronomy* 14, no. 3 (1906): 129–35.

Hussey, W.J. 'The Lick Observatory–Crocker Eclipse Expedition to Egypt.' *Publications of the Astronomical Society of the Pacific* 18 (1906): 37–46.

Ibn Bāz, 'Abd al-'Azīz. *Al-adillah al-naqlīyah wa-ḥissīyah 'alá imkān al-ṣu'ūd ilá al-kawākib wa-'alá jiryān al-shams wa-al-Qamar wa-sukūn al-Arḍ*. Al-ṭab'ah al-thāniyah. Al-Riyāḍ: Maktabat al-Riyāḍ al-Ḥadīthah, 1982.

———. 'Takdhīb wa-naqd mā nasharat'hu majallat (al-Muṣawwar).' *Al-mawqi' al-rasmī li-samāḥat al-imām Ibn Bāz*. Accessed 28 October 2016. http://www.binbaz.org.sa/article/403.

———. 'Radd ʿalá al-muftarīn ʿalá al-ʿulamāʾ.' *Al-mawqiʿ al-rasmī li-samāḥat al-imām Ibn Bāz*. Accessed 28 October 2016. http://www.binbaz.org.sa/article/472.

———. 'Ḥukm al-qawl bi-anna awwal rāʾid faḍāʾ nabīyinā Muḥammad ṣallá Allāh ʿalayhi wa-sallam.' *Al-mawqiʿ al-rasmī li-samāḥat al-imām Ibn Bāz*. Accessed 28 October 2016. http://www.binbaz.org.sa/noor/8684.

Ibrahim, Makram, Y.S. Hanna, S.W. Samwel, and Maroof A. Hegazy. 'Satellite Laser Ranging in Egypt.' *NRIAG Journal of Astronomy and Geophysics* 4, no. 1 (2015): 123–29.

International Astronomical Union. 'With the Exhibits.' *IAU News Bulletin* 3 (17 August 1961): 6–7.

Issa, I.A., and A.I. Gamal El Din. 'Astronomical Research Activities with the 74 Inch Telescope at Kottamia Observatory.' *Astrophysics and Space Science* 118 (1986): 87–93.

Issa, Laudy. 'Petition Circulates to Turn Lee Observatory into Space Museum.' *Outlook* 49, no. 8 (2016): 1–2.

Jalal, Maán. 'These Guys Know Their Aliens.' *Khaleej Times*, 26 June 2016. http://www.khaleejtimes.com/these-guys-know-their-aliens.

Johnson, Alice. 'Science fiction for young adults expands in the UAE.' *The National*, 5 March 2012. http://www.thenational.ae/arts-culture/books/science-fiction-for-young-adults-expands-in-the-uae.

Jones, Sophia. 'Top Egyptian Comedian Slams Military for Saying It Can "Cure" AIDS and Hepatitis C.' *Huffington Post*, 3 March 2014. http://www.huffingtonpost.com/2014/03/03/bassem-youssef-aids-hepatitis_n_4889382.html.

Joseph, Bob. 'Faculty Profile: Shadia Rifai Habbal.' *Nā Kilo Hōkū: A Newsletter from the Institute of Astronomy, University of Hawaiʻi* 18 (2006). http://www2.ifa.hawaii.edu/newsletters/article.cfm?a=263&n=1.

Journal of the Royal Central Asian Society. 'Obituary: Faris Nimr Pasha.' *Journal of the Royal Central Asian Society* 39, no. 2 (1952): 167–69.

Joy, Alfred H. 'Transit of Mercury, Nov. 13–14, 1907.' *Popular Astronomy* 16 (1908): 128.

———. 'Photometric measures of BD +13°4692.' *Astronomische Nachrichten* 200, no. 1 (1914): 11–12.

Jurdak, Mansur Ḥanna. *Astronomical Dictionary: The Zodiac & the Constellations*. Beirut: American Mission Press, 1950.

Karam, Michael. 'Orderly fashion not the style in Beirut.' *The National*, 28 September 2015. http://www.thenational.ae/business/the-life/orderly-fashion-not-the-style-in-beirut.

Kellermann, K.I. *David S. Heeschen, 1926–2012: A Biographical Memoir*. National Academy of Sciences, 2014.

Kelley, Kevin W. *The Home Planet*. Reading: Addison-Wesley, 1988.

Khaleej Times. 'Khalifa Sat to be ready for 2017 launch: Mohammed.' *Khaleej Times*, 31 December 2013. http://www.khaleejtimes.com/nation/government/khalifa-sat-to-be-ready-for-2017-launch-mohammed.

Khoury, Jack. 'Osama Al Baz, an architect of Israel–Egypt peace, dies at 82.' *Haaretz*, 17 September 2013. http://www.haaretz.com/news/middle-east/.premium-1.547379.

King, David A. 'Smithsonian Institution Project in Medieval Islamic Astronomy.' *Historia Mathematica* 1, no. 2 (1974): 183–84.

———. 'The Astronomy of the Mamluks.' *Isis* 74, no. 4 (1983): 531–55.

———. 'Frans Bruin (1922–2001).' *Journal for the History of Astronomy* 33 (2002): 214–16.

———. *In Synchrony with the Heavens: Studies in Astronomical Timekeeping and Instrumentation in Medieval Islamic Civilization.* Vol. 2. Leiden: Brill, 2005.

King, David A., and Owen Gingerich. 'Some Astronomical Observations from Thirteenth-Century Egypt.' *Journal for the History of Astronomy* 13 (1982): 121–28.

King Abdulaziz University. 'Former Distinguished Adjunct Professors.' *KAU*, 12 February 2015. http://dsc.kau.edu.sa/Pages-Former-Distinguished-Adjunct-Professors.aspx.

Kissinger, Henry. 'Smithsonaian /SFCP-Further Disbursement Unde Grant SFG-2-8464 Theories of Planetary Motion.' *WikiLeaks*, 27 December 1973. https://www.wikileaks.org/plusd/cables/1973STATE250820_b.html.

———. 'Visit by Astronaut Crew.' *WikiLeaks*, 16 January 1976. https://wikileaks.org/plusd/cables/1976STATE010856_b.html.

Klārk, Arthar [see also Clarke, Arthur]. 'Sa-nusāfir ilá al-Qamar.' *Al-Hilāl* 58, no. 1 (1950): 97–100.

Kobold, H. 'Mitteilungen über die Nova Aquilae 3.' *Astronomische Nachrichten* 207, no. 18 (1918): 193–204.

Kopal, Zdenek. Letter to J.C. Clark, 20 May 1964. Papers of Zdenek Kopal, GB 133 KOP. John Rylands Library, University of Manchester.

———. Letter to A.H. Samaha, 10 March 1967. Papers of Zdenek Kopal, GB 133 KOP. John Rylands Library, University of Manchester.

———. Letter to A.H. Samaha, 21 December 1967. Papers of Zdenek Kopal, GB 133 KOP. John Rylands Library, University of Manchester.

———. Letter to A.S. Asaad, 17 September 1968. Papers of Zdenek Kopal, GB 133 KOP. John Rylands Library, University of Manchester.

———. 'Preface.' *Astrophysics and Space Science* 1, no. 1 (1968): 3–5.

———. 'John M. Saari.' *Moon* 2 (1971): 405–7.

———. *Of stars and men: Reminiscences of an astronomer*. Bristol: IOP, 1986.

Kuwait News Agency. 'Marṣad al-ʿUjayrī al-Falakī. sittat ʿashar ʿāman min al-taṭawwur wa-al-ʿaṭāʾ al-ʿilmī.' *Kuwait News Agency*, 13 April 2002. http://www.kuna.net.kw/ArticlePrintPage.aspx?id=1244922.

Lawton, John, and Patricia Moody. 'A Prince in Space.' *Aramco World* 37, no. 1 (1986): 20–29.

Laylin, Tafline. 'NASA Made a Crazy Planet Discovery. This Is the 25-Year-Old Moroccan Who Helped.' *OZY*, 17 May 2017. http://www.ozy.com/rising-stars/meet-the-moroccan-scientist-who-helped-find-7-earth-like-exoplanets/76636.

Lobel, Mark. 'The United Arab Emirates reaches for the Red Planet.' *BBC News*, 29 May 2015. http://www.bbc.com/news/business-32919988.

Loomis, Ilima. 'This Hawai'i Astronomer Chases Solar Eclipses Around the World Trying to Solve the Biggest Mysteries of the Sun.' *Honolulu*, 21 September 2015. http://www.honolulumagazine.com/Honolulu-Magazine/September-2015/This-Hawaii-Astronomer-Chases-Solar-Eclipses-Around-the-World-Trying-to-Solve-the-Biggest-Mysteries-of-the-Sun/.

Lyot, Bernard, and Audouin Dollfus. 'Étude des spectres de la couronne solaire obtenus à Khartoum pendant l'éclipse totale de Soleil du 25 février 1952.' *Comptes rendus hebdomadaires des séances de l'Académie des sciences* 237, no. 16 (1953): 855–59.

Madwar, M.R. *Preliminary Notice on the Total Solar Eclipse of 25th February 1952*. Cairo: Fouad I University Press, 1952.

———. 'Solar Work and Plans in Egypt.' *Vistas in Astronomy* 1 (1955): 732–37.

Mahoney, Lynn. 'Welcome to the "New" West Hall.' *Main Gate: American University of Beirut Quarterly Magazine* 1, no. 3 (2003).

Makiya, Kanan. *Republic of Fear: The Politics of Modern Iraq*. Updated Edition. Berkeley: University of California Press, 1998.

Malek, Caline. 'UAE's Mars space mission has a new name: Hope.' *The National*, 6 May 2015. http://www.thenational.ae/uae/uaes-mars-space-mission-has-a-new-name-hope.

Mahmoud Effendi. 'Mémoire sur le Calendrier arabe avant l'islamisme, et sur la naissance et l'âge du prophète Mohammad.' *Journal asiatique*, 5th series, 11 (1858): 109–91.

Malawi, Abdulrahman Ali. 'Atomic hydrogen associated with high latitude IRAS cirrus clouds.' PhD thesis, University of Manchester, 1989.

Markaz Qaṭar li-ʿUlūm al-Faḍāʾ wa-al-Falak. 'Man naḥnu.' *Markaz Qaṭar il-ʿUlūm al-Faḍāʾ wa-al-Falak*. Accessed 14 May 2015. http://qatar-falak.net/We-are.aspx.

Masevich, A.G., and K. Hamal. 'Intercosmos Laser Ranging Stations.' In *Space Research XVII*, edited by M.J. Rycroft and A.C. Strickland, 73–76. Oxford: Pergamon, 1977.

McNally, Derek, *ed. Proceedings of the Twentieth General Assembly, Baltimore 1988*. Dordrecht: Kluwer, 1990.

———. '(10478) Alsabti =1981 WO.' *Minor Planet Circulars*, 1 November 2001: 43762.

Meaburn, John. 'Zdenek Kopal (1914–1993).' *Quarterly Journal of the Royal Astronomical Society* 35 (1994): 229–30.

Memory Alpha. 'El-Baz.' *Memory Alpha, the Star Trek Wiki*. Accessed 23 November 2013. http://en.memory-alpha.org/wiki/El-Baz.

———. 'USS Farouk El-Baz.' *Memory Alpha, the Star Trek Wiki*. Accessed 23 November 2013. http://en.memory-alpha.org/wiki/USS_Farouk_El-Baz.

Mikhail, Joseph Sidky. 'Color Opposition Effect of the Lunar Surface.' *Icarus* 8, no. 1 (1968): 117–28.

———. 'Telescopic Investigation of Relative Spectral Reflectivity of Mare Grounds of Lunar Surface.' *Moon and the Planets* 20, no. 2 (1979): 199–210.

Mikhail, Joseph Sidky, and I.K. Koval. 'Dependence of Normal Albedo of Lunar Regions on Wavelength.' *Moon* 11 (1974): 323–26.

Mīlād, Jirjis. 'Shāhid. Fārūq al-Bāz yakshif tafāṣīl liqā'ihi bi-al-ra'īs al-Sīsī.' *Al-Wafd*, 15 October 2016.

Miley, George. 'Saudi University Policy: Overvalued Rankings.' *Science* 335, no. 6072 (2012): 1041–42.

Miller, Judith. 'In Saudi Arabia, the Sun Shines Bright on Solar Power.' *New York Times*, 1 November 1983.

Millman, P.M. 'Working Group for Planetary System Nomenclature.' In *Proceedings of the Sixteenth General Assembly, Grenoble 1976*, edited by Edith A. Müller and Arnost Jappel, 321–69. Dordrecht: D. Reidel, 1977.

Mimouni, Jamal. 'Popularising science through astronomy, an Algerian experience in grassroot activism and its academic spin-off.' In *The Role of Astronomy in Society and Culture*, edited by David Valls-Gabaud and Alexander Boksenberg, 741–47. Cambridge: Cambridge University Press, 2011.

Mimouni, Jamal, and Nassim Seghouani. 'L'Ecole Doctorale d'Astrophysique.' *La lettre du CRAAG*, no. 39 (2008): 4–5.

Mislis, D., E. Bachelet, K.A. Alsubai, D.M. Bramich, and N. Parley. 'SIDRA: a blind algorithm for signal detection in photometric surveys.' *Monthly Notices of the Royal Astronomical Society* 455, no. 1 (2016): 626–33.

Mislis, D., S. Pyrzas, K.A. Alsubai, Z.I. Tsvetanov, and N.P.E. Vilchez. 'The DOHA algorithm: a new recipe for cotrending large-scale transiting exoplanet survey light curves.' *Monthly Notices of the Royal Astronomical Society* 465, no. 3 (2017): 3759–66.

Morgan, Michael Hamilton. *Arab Science: A Journey of Innovation*. Doha: Qatar Foundation, 2009.

Morrison, David. 'IAU Nomenclature for Topographic Features on Mercury.' *Icarus* 28, no. 4 (1976): 605–06.

Muftāḥ, Maḥmūd ʿAbd al-Raḥmān. *Al-samā' wa-al-aṭbāq al-ṭā'irah*. Al-Dawḥah: ʿAlī ibn ʿAlī, 1985.

Mujāhid, Imād. 'Al-Shaykh Salmān Āl Thānī: iktishāf fawhah nayzakīyah ḥadīthat al-takwīn fī Qaṭar.' *Al-Dustūr*, 30 March 2008.

Müller, Edith A., and Arnost Jappel, eds. *Proceedings of the Sixteenth General Assembly, Grenoble 1976*. Transactions of the International Astronomical Union XVIB. Dordrecht: Reidel, 1977.

Muṣṭafá, Nūrhān. 'Tārīkh ashhar aḥyā' al-Qāhirah mundh al-nash'ah ḥattá al-ān: 'Bāb al-Lūq' qalb waṣat al-balad (ḥ 2).' *Al-Maṣrī al-Yawm*, 14 December 2015. http://lite.almasryalyoum.com/lists/79695/.

Nagel, Steven R., and Jennifer Ross-Nazzal. 'Oral History Transcript.' *NASA Johnson Space Center Oral History Project*, 20 December 2002. http://www.jsc.nasa.gov/history/oral_histories/NagelSR/NagelSR_12-20-02.pdf.

Naifeh, George. 'Astronauts Visit.' *WikiLeaks*, 15 February 1976. https://wikileaks.org/plusd/cables/1976ABUDH00381_b.html.

Nather, R.E., D.E. Winget, J.C. Clemens, C.J. Hansen, and B.P. Hine. 'The Whole Earth Telescope: A New Astronomical Instrument.' *Astrophysical Journal* 361 (1990): 309–17.

National Academy of Sciences. *Report of the National Academy of Sciences: Fiscal Year 1924–1925*. Washington: Government Printing Office, 1926.

Neufeld, Michael J. 'The Nazi aerospace exodus: towards a global, transnational history.' *History and Technology* 28, no. 1 (2012): 49–67.

New Scientist. 'Reign of terror sweeps Iraq's scientific community.' *New Scientist* 90, no. 1247 (2 April 1981): 3–4.

New York Post. 'Iraq to release Saddam Hussein's last WMD scientist.' *New York Post*, 15 April 2012. http://nypost.com/2012/04/15/iraq-to-release-saddam-husseins-last-wmd-scientist/.

Nimr, Fāris. 'Baʿda ʿahdī bi-ʿilm al-falak.' *Al-Muqtaṭaf* 92, no. 5 (1938): 483–89.

Novotny, Antonin, and Josef Blazej. 'Human factor as an important phenomenon for the optical radar station operation.' In *15th Czech–Polish–Slovak Conference on Wave and Quantum Aspects of Contemporary Optics*, edited by Miroslav Miler, Dagmar Senderáková, and Miroslav Hrabovský. SPIE, 2007.

Nuʿmān, Shayth. 'Muṣṭalaḥāt fī ʿulūm al-faḍāʾ.' *Majallat al-Majmaʿ al-ʿIlmī al-ʿIrāqī*, no. 6 (March 1959): 355–67.

Office of Technology Assessment. *UNISPACE '82: A Context for International Cooperation and Competition: A Technical Memorandum*. Washington: Government Printing Office, 1983.

Okaz. 'Awwal rāʾid faḍāʾ Sūrī. al-niẓām yukāfiʾ ʿulamāʾahu bi-al-tajāhul.' *Okaz*, 3 September 2012. http://www.okaz.com.sa/new/Issues/20120903/Con20120903529603.htm.

O'Neill, Katherine Jane. 'How Mursi's English "destroyed" his American students.' *Al Arabiya*, 10 April 2013. http://ara.tv/2396p.

Oosterhoff, P. Th., ed. *Ninth General Assembly Held at Dublin, 29 August to 5 September 1955*. Cambridge: Cambridge University Press, 1957.

Osama, Athar. 'Profiles in Leadership (4): Dr Farouk El-Baz on Apollo Programme, Egyptian Science, and Desert Development Corridor.' *Muslim Science*, 19 July 2011. http://muslim-science.com/profiles-in-leadership-series-4-dr-farouk-el-baz-on-apollo-programme-egyptian-science-and-desert-development-corridor/.

Paganelli, Robert P. 'Visit of Astronaut Colonel James Irwin to Qatar.' *WikiLeaks*, 17 December 1974. https://wikileaks.org/plusd/cables/1974DOHA00037_b.html.

———. 'Smithsonian Activities in Qatar: Visit of Dr. Farouk El Baz.' *WikiLeaks*, 11 January 1975. https://wikileaks.org/plusd/cables/1975DOHA00034_b.html.

Peacock, Keith. 'Multicolor Photoelectric Photometry of the Lunar Surface.' *Icarus* 9, no. 1–3 (1968): 16–66.

Permanent Delegation of the Arab Republic of Egypt to Unesco. 'Helwan Observatory.' *UNESCO World Heritage Centre*, 11 March 2010. http://whc.unesco.org/en/tentativelists/5574/.

Physics Today. 'Julius Arthur Brown.' *Physics Today* 23, no. 9 (1970): 83.

Plackett, Benjamin. 'A Conversation with Farouk El-Baz: Finding Water in the Desert.' *Al-Fanar Media*, 10 September 2014. http://www.al-fanarmedia.org/2014/09/conversation-farouk-el-baz-finding-water-desert/.

Pollacco, D., I. Skillen, A. Collier Cameron, B. Loeillet, H.C. Stempels, F. Bouchy, N.P. Gibson, et al., 'WASP-3b: a strongly irradiated transiting gas-giant planet.' *Monthly Notices of the Royal Astronomical Society* 385, no. 3 (2008): 1576–84.

Qadrī, Ḥusayn. *Hurūb ilá al-faḍā'*. Al-Qāhirah: Dār al-Ma'ārif, 1983.

Qassūm, Niḍāl [see also Guessoum, Nidhal], Jamāl Maymūnī, and Karīm Muzyān. ''Ilm al-falak fī al-Jazā'ir: māḍiyan, ḥāḍiran, mustaqbalan.' *Association Syrius d'Astronomie*, 1999. http://siriusalgeria.net/astroinalgarabic.htm.

Qatar Foundation. 'Contents.' *The Foundation*, no. 26 (February 2011): 1.

———. 'QF discovers another planet: Qatar-2b.' *Qatar Foundation Telegraph*, no. 37 (2011): 1–2.

Qatar is Booming. 'Stars of Science Series 3 – The Competition Begins.' *Qatar is Booming*, 2 November 2011. http://www.qatarisbooming.com/2011/11/02/stars-of-science-series-3-the-competition-begins/.

Qatar National Research Fund. 'Khalid Al Subai: Searching for New Worlds.' *QNRF Newsletter*, no. 12 (August 2013). http://qnrfnewsletter.org/issue12/interview.php.

Qubaysī, Kamāl. 'Miṣrī ikhtārūhu lil-safar ilá al-Mirrīkh sa-yazra' fīhi al-mulūkhīyah.' *Al Arabiya*, 23 February 2015. http://ara.tv/mrsvk.

Querci, François R. 'Network of Oriental Robotic Telescopes.' In *Highlights of Astronomy as Presented at the XXIInd General Assembly of the IAU, 1994*, edited by I. Appenzeller, 677–79. Dordrecht: Kluwer, 1995.

Querci, Francois Rene, Monique Querci, Samir Kadiri, and Zouhair Benkhaldoun. 'Network of Oriental Robotic Telescopes.' In *Robotic Telescopes: Current Capabilities, Present Developments, and Future Prospects for Automated Astronomy*, edited by Gregory W. Henry, Gregory W. Eaton, and Joel A. Eaton, 239–41. San Francisco: Astronomical Society of the Pacific, 1995.

Querci, F.R., M. Querci, S. Kadiri, and L. de Rancourt. 'A complementary network to GNAT: An Arabian and French project for 3T1M automated photometric stations.' In *Robotic Observatories*, edited by Michael F. Bode, 85–88. Chichester: Wiley, 1995.

Querci, Francois R., and Monique Querci. 'High Precision Photometry: An Automated Station Project with Three 1-M Telescopes.' In *Automated Telescopes for Photometry and Imaging*, edited by Saul J. Adelman, Robert J. Dukes, and Carol J. Adelman, 67–71. San Francisco: Astronomical Society of the Pacific, 1992.

Qunṣul, Khalīl. ''Askarat al-faḍā': istrātījīyat al-tawattur al-dā'im wa-al-ru'yah al-Amrīkīyah li-ṣiyāghat al-ʿālam.' *Al-Majallah al-Thaqāfīyah*, no. 60 (2003): 311–46.

Qunṣul, Khalīl, and Ḥamīd Mujwil al-Naʿīmī. *Taqrīr nashāṭāt al-Ittiḥād al-ʿArabī li-ʿUlūm al-Falak wa-al-Faḍā'*. Amman: Arab Union for Astronomy and Space Sciences, 2002.

RAND. 'Farouk El-Baz Joins RAND-Qatar Policy Institute Board.' *RAND*, 7 June 2004. http://www.rand.org/news/press/2004/06/07.html.

Ratney, Michael. 'Qatari Amir's Wife Visits Iraq, Furthers Political, Educational Cooperation.' *WikiLeaks*, 11 May 2009. https://wikileaks.org/plusd/cables/09DOHA311_a.html.

Raydan, Noam. 'No cure for pseudoscience.' *NOW*, 22 April 2014. https://now.mmedia.me/lb/en/Interview/544209-talking-to-mohammad-yahia.

Republic of Iraq. *Iraqi Academic Conference to Help Improve and Sustain Higher Education in Iraq*. Washington: Embassy of the Republic of Iraq, Cultural Office, 2009.

Rifaat, A.S. 'Photometric Studies of Two Lunar Domes.' *Icarus* 7, no. 1 (1967): 267–73.

Riley, Charles. 'Star Wars puts Abu Dhabi on Hollywood's map.' *CNN*, 15 December 2015. http://money.cnn.com/2015/12/15/media/star-wars-abu-dhabi-film-tax-credits/index.html.

Rockefeller Foundation. *Annual Report, 1957*. New York: Rockefeller Foundation, 1958.

Royal Astronomical Society. 'Meeting of the Royal Astronomical Society.' *Observatory* 46, no. 585 (1923): 33–41.

Russell, Henry Norris. 'Raymond Smith Dugan.' *Popular Astronomy* 48 (1940): 466–69.

Saad, Rehab. 'A Writer and Diplomat: Abdel-Hamid Abdel-Ghani (Abdel-Hamid El-Katib) (1917–2000).' *Al-Ahram Weekly* 465, 20–26 January 2000. http://weekly.ahram.org.eg/2000/465/eg9.htm.

Saari, J.M., and R.W. Shorthill. 'Isotherms of Crater Regions on the Illuminated and Eclipsed Moon.' *Icarus* 2 (1963): 115–36.

———. 'Thermal Anomalies on the Totally Eclipsed Moon of December 19, 1964.' *Nature* 205, no. 4975 (1965): 964–65.

Saari, J.M., R.W. Shorthill, and T.K. Deaton. 'Infrared and Visible Images of the Eclipsed Moon of December 19, 1964.' *Icarus* 5, no. 1 (1966): 635–59.

Saʿd, Sārah. 'Marṣad Jāmiʿat al-Malik ʿAbd al-ʿAzīz al-waḥīd Khalījīyan wa-al-thānī ʿArabīyan.' *Al-Yawm*, 28 December 2011. http://www.alyaum.com/article/3039386.

Saied, Mohamed. 'Is there hope for Egypt's first space odyssey?' *Al-Monitor*, 4 June 2015. http://www.al-monitor.com/pulse/originals/2015/06/egypt-space-agency-sisi-draft-law-heggy-narss.html.

Samaha, Abdel Hamid. 'Notes as to cosmological ideas in Al Quran.' *Meddelanden fran Lunds Astronomiska Observatorium II*, no. 96 (1938): 3–5.

———. 'The Kottamia 74' Telescope.' *Helwan Observatory Bulletin* 62 (1964).

———. Letter to Z. Kopal, 28 February 1967. Papers of Zdenek Kopal, GB 133 KOP. John Rylands Library, University of Manchester.

———. 'Lunar Research at the Kottamia Observatory.' In *Measure of the Moon*, edited by Zdeněk Kopal and Constantine L. Goudas, 78–88. Dordrecht: Reidel, 1967.

Samaha, A.H., and Z. Kopal. 'Lunar Photography with the 74-Inch Reflector of the Helwan Observatory at Kottamia.' *Icarus* 5 (1966): 309–12.

Sardar, Ziauddin. 'Middle East brain drain switches back from the West.' *Nature* 283, no. 5745 (1980): 327–28.

———. 'Shake-up in Syrian science imminent.' *New Scientist* 87, no. 1208 (3 July 1980): 6.

Ṣarrūf, Fu'ād. 'Qīmat al-baḥth al-'ilmī wa-muhimmat al-Muqtaṭaf.' *Al-Muqtaṭaf* 69, no. 1 (1926): 71–75.

———. 'Naḥnu shu'ūb al-'ālam.' *Al-Adīb*, no. 4 (1 April 1950): 3–4.

———. 'Ghazw al-faḍā'.' *'Ālam al-Fikr* 1, no. 3 (1970): 9–42.

Ṣarrūf, Ya'qūb, and Fāris Nimr. 'Muqaddimah.' *Al-Muqtaṭaf* 1, no. 1 (1876): 1–2.

———. ''Ulamā' al-hay'ah 'ind al-'Arab.' *Al-Muqtaṭaf* 1, no. 1 (1876): 16–18.

———. 'Min al-marṣad al-Sūrī al-falakī wa-al-mitiyūrūlūjī.' *Al-Muqtaṭaf* 1, no. 2 (1876): 46–47.

———. ''Ubūr al-Zahrah.' *Al-Muqtaṭaf* 7, no. 5 (1882): 308.

———. ''Ubūr al-Zahrah wa-bu'd al-shams.' *Al-Muqtaṭaf* 7, no. 6 (1883): 321–26.

———. 'Al-kusūf al-muqbil.' *Al-Muqtaṭaf* 30, no. 4 (1905): 257–59.

Schaufele, William E. 'UN Secretariat Staffing.' *WikiLeaks*, 26 April 1974. https://wikileaks.org/plusd/cables/1974USUNN01540_b.html.

Schenectady Gazette. 'Albany State Professor Wins $36,419 From NSF.' *Schenectady Gazette*, 5 September 1969: 25.

Schreck, Adam. 'Abu Dhabi partners with Virgin Galactic spaceship firm.' *ABC News*, 28 July 2009. http://abcnews.go.com/Business/story?id=8191703&page=1.

Sciences et avenir. 'Le colonel Kadhafi s'offre un télescope.' *Sciences et avenir*, 4 January 2005. http://www.sciencesetavenir.fr/espace/le-colonel-kadhafi-s-offre-un-telescope_32538.

Shaaban, Ahmed. 'One-way trip to Mars prohibited in Islam.' *Khaleej Times*, 19 February 2014. http://www.khaleejtimes.com/nation/general/one-way-trip-to-mars-prohibited-in-islam.

Shahdā, 'Āmir. 'Al-Suwayd tarfuḍ manḥ awwal rā'id faḍā' Sūrī fīsā li-ḥuḍūr mu'tamar duwalī." *Orient Net*, 10 September 2015. http://www.orient-news.net/ar/news_show/90492.

Shaheen, Kareem. 'Emirates Space Mission Hopes to Launch New Era in Middle East.' *The Guardian*, 19 July 2015, http://www.theguardian.com/world/2015/jul/19/emirates-space-mars-mission-middle-east.

Shaltout, M.A.M [see also Shaltūt, Musallam]. 'The Marskhod Egyptian Drill Project.' In *Developing Basic Space Science World-Wide: A Decade of UN/ESA*

*Workshops*, edited by Willem Wamsteker, Rudolf Albrecht, and Hans J. Haubold, 241–49. Dordrecht: Kluwer, 2004.

———. 'Small Astronomical Telescopes for Research and Education at Helwan, Egypt.' In *Developing Basic Space Science World-Wide: A Decade of UN/ESA Workshops*, edited by Willem Wamsteker, Rudolf Albrecht, and Hans J. Haubold, 251–59. Dordrecht: Kluwer, 2004.

Shaltūt, Musallam [see also Shaltout, M.A.M.]. 'Min al-ishārāt al-kawnīyah fī al-Qur'ān al-karīm. wa-al-samā' dhāt al-raj'.' *Al-I'jāz al-'Ilmī*, no. 17 (2004): 10–15.

Sharma, Suraj. 'Syrian in space: From cosmonaut hero to refugee.' *Middle East Eye*, 3 December 2015. http://www.middleeasteye.net/in-depth/features/muhammed-faris-space-odyssey-refugee-odyssey-1880597043.

Shiha, Gamal, Waleed Samir, Zahid Azam, Premashis Kar, Saeed Hamid, and Shiv Sarin. 'A Novel Method for Non-Invasive Diagnosis of Hepatitis C Virus Using Electromagnetic Signal Detection: A Multicenter International Study.' *International Journal of Medical, Health, Biomedical, Bioengineering and Pharmaceutical Engineering* 7, no. 12 (2013): 996–1003.

Shorthill, R.W., and J.M. Saari. 'Nonuniform Cooling of the Eclipsed Moon: A Listing of Thirty Prominent Anomalies.' *Science* 150, no. 3693 (1965): 210–12.

Silbermann, Paul. 'Historical Note.' *Bellcomm, Inc Technical Library Collection*, 2001. http://airandspace.si.edu/research/arch/findaids/bellcomm/bci_sec_3.html.

Simmons, Mike. 'Seeing Stars in Iraq.' *Scientific American* 296, no. 1 (2007): 23–25.

Space Daily. 'Egypt Mulls Buying Russian Satellite Images After EgyptSat 2 Loss.' *Space Daily*, 9 June 2015. http://www.spacedaily.com/reports/Egypt_Mulls_Buying_Russian_Satellite_Images_After_EgyptSat_2_Loss_999.html.

SpaceWatch Middle East. 'China To Contribute Funds to Egypt's Space Programm.' *SpaceWatch Middle East*, March 2017. https://spacewatchme.com/2017/03/china-contribute-funds-egypts-space-programme/.

———. 'UAE's MBRSC Reveals Further Details About Its Mars 2117 Colonisation Programme.' *SpaceWatch Middle East*, 11 May 2017. https://spacewatchme.com/2017/05/uaes-mbrsc-reveals-details-mars-2117-colonisation-programme/.

Spitzer Science Center. 'Astronomical Success: Caltech's George Helou Honored by Home Country of Lebanon.' *Caltech*, 25 January 2012. http://www.spitzer.caltech.edu/news/1364-feature12-02-Astronomical-Success-Caltech-s-George-Helou-Honored-by-Home-Country-of-Lebanon-.

St. Cyr, O.C., Joseph M. Davila, Madhulika Guhathakurta, and Robert S. Senseney. 'Science Eclipses Politics During 2006 U.S. Eclipse Expedition to Libya.' *Eos* 88, no. 49 (2007): 539–40.

Storey, Kenneth. 'Question Time.' *Aramco World* 37, no. 1 (1986): 34–35.

Storey, Kenneth, and Joseph Fitchett. 'Down to Earth.' *Aramco World* 37, no. 1 (1986): 30–40.

Struve, Otto. Letter to M.H. Jurdak, 8 November 1954. Mansour Hanna Jurdak Collection, 1881–1964, AA:6.2.10, Box no. 1, File no. 6. American University of Beirut/Library Archives.

———. Letter to M. Jurdak, 3 March 1955. Mansour Hanna Jurdak Collection, 1881–1964, AA:6.2.10, Box no. 1, File no. 6. American University of Beirut/Library Archives.

Subcommittee on Space Science and Applications. *Report on UNISPACE '82: The United Nations Conference on the Exploration and Peaceful Uses of Outer Space, Vienna, Austria, August 9–21, 1982*. Washington: House of Representatives, 1982.

Sultan, A'haq, and François René Querci. 'The Mountains of Yemen: The Most Suitable Location for a Regional Observatory.' In *Seminars of the United Nations Programme on Space Applications: Selected Papers from Activities Held in 2001*, 71–76. New York: United Nations, 2002.

Sultan, Zainab. 'Glittering prize for Qatar's planet hunter.' *Nature Middle East*, 13 May 2013. http://www.nature.com/nmiddleeast/2013/130513/full/nmiddleeast.2013.71.html.

Thomas, Selin. 'Dark Reality Catches Up to a Show About a Syrian Astronaut.' *Village Voice*, 31 January 2017. http://www.villagevoice.com/arts/dark-reality-catches-up-to-a-show-about-a-syrian-astronaut-9624742.

Todd, D.P. 'The Solar Parallax as derived from the American Photographs of the Transit of Venus, 1874, December 8–9.' *The Observatory* 4 (1881): 202–05.

———. 'The Amherst Eclipse Expedition to Singkep, 1901.' *Astrophysical Journal* 14 (1901): 362–66.

———. 'The Singkep Mechanical Commutator.' *Journal of the British Astronomical Association* 12, no. 4 (1902): 169–75.

———. 'Early History of Astronomy at Amherst College.' *Popular Astronomy* 11 (1903): 322–26.

———. 'Observations of the Total Solar Eclipse of 1905, August 30, at Tripoli, Barbary.' *Science* 23, no. 586 (1906): 462–63.

Todd, Mabel Loomis. *Tripoli the Mysterious*. London: Grant Richards, 1912.

True. 'Rā'id al-faḍā' Muḥammad Fāris fī mukālamah maʿa al-raʾīs Ḥāfiẓ al-Asad 1987.'*YouTube*, 22 July 2011. https://www.youtube.com/watch?v=fiyY0T-Zou7w.

Turner, Herbert Hall. 'From an Oxford Note-Book.' *Observatory* 28 (1905): 395–400.

Turner, H.H., and H.F. Newall. 'Total Solar Eclipse of 1900 (May 28). Preliminary Report on the Observations Made at Bouzareah (in the Grounds of the Algiers Observatory).' *Proceedings of the Royal Society of London* 67 (1900): 346–69.

Ulmer, M.P., R.G. Cruddace, G.G. Fritz, W.A. Snyder, and E.E. Fenimore. 'Spartan 1 X-Ray Observations of the Perseus Cluster: Comparison of the

Iron Abundances and Temperatures in the Inner and Outer Regions of the Clusters.' *Astrophysical Journal* 319 (August 1987): 118–25.
'Umrān, Ṭālib. *Faḍā' wāsi' ka-al-ḥulm*. Dimashq: Dār al-Fikr, 1997.
———. 'Kutlah ḥayyah bi-ḥajm kawkab.' *Al-Khayāl al-'Ilmī*, no. 27 (2010): 28–39.
———. 'Ruwwād al-Kawkab al-Aḥmar 1/2.' *Al-Adab al-'Ilmī*, no. 14 (2014): 112–21.
———. 'Faḍā' wāsi' ka-al-ḥulm 1/2.' *Al-Adab al-'Ilmī*, no. 16 (2014): 110–31.
United Nations Development Programme. *The Arab Human Development Report 2005: Towards the Rise of Women in the Arab World*. New York: United Nations, 2006.
United Nations General Assembly. '1348 (XIII). Question of the peaceful use of outer space.' *UNOOSA*, 13 December 1958. http://www.oosa.unvienna.org/oosa/SpaceLaw/gares/html/gares_13_1348.html.
———. '1472 (XIV). International co-operation in the peaceful uses of outer space.' *UNOOSA*, 12 December 1959. http://www.unoosa.org/oosa/SpaceLaw/gares/html/gares_14_1472.html.
———. '1721 (XVI). International co-operation in the peaceful uses of outer space.' *UNOOSA*, 20 December 1961. http://www.oosa.unvienna.org/oosa/SpaceLaw/gares/html/gares_16_1721.html.
———. '2221 (XXI). United Nations Conference on the Exploration and Peaceful Uses of Outer Space.' 19 December 1966. Twenty-first session. United Nations General Assembly Official Records.
———. '2222 (XXI). Treaty on Principles Governing the Activities of States in the Exploration and Use of Outer Space, Including the Moon and Other Celestial Bodies.' *UNOOSA*, 19 December 1966. http://www.unoosa.org/oosa/en/ourwork/spacelaw/treaties/outerspacetreaty.html.
———. '32/196. International co-operation in the peaceful uses of outer space.' *UNOOSA*, 20 December 1977. http://www.unoosa.org/oosa/en/SpaceLaw/gares/html/gares_32_0196.html.
———. '35/16. Enlargement of the Committee on the Peaceful Uses of Outer Space.' *UNOOSA*, 3 November 1980. http://www.unoosa.org/oosa/SpaceLaw/gares/html/gares_35_0016.html.
United Nations Monitoring, Verification and Inspection Commission. *Compendium of Iraq's Proscribed Weapons Programmes in the Chemical, Biological and Missile Areas*. United Nations, 2007.
Valdini, Claire. 'Dr Khalid Al Subai Interview.' *Arabian Business*, 24 June 2012. http://www.arabianbusiness.com/best-of-2012-dr-khalid-al-subai-interview-462982.html.
Van Aerschodt, L. 'Raymond Smith Dugan (1878–1940).' *Ciel et Terre* 57 (1941): 308–09.
Vān Dayk, Kurnīliyūs. *Uṣūl 'ilm al-hay'ah*. Bayrūt, 1874.
Van Dyck, C.V.A. 'On the Present Condition of the Medical Profession in Syria.' *Journal of the American Oriental Society* 1, no. 4 (1849), 559–91.

Van Maanen, A., J.A. Brown, and M.L. Humason. 'A Star of Extremely Low Luminosity.' *Publications of the Astronomical Society of the Pacific* 39 (1927): 173–74.

Varghese, Joseph. 'Qatar to set up a space centre.' *Gulf Times*, 11 February 2013. http://www.gulf-times.com/qatar/178/details/341759/qatar-to-set-up-a-space-centre.

Viéville, T., G. Clément, F. Lestienne, and A. Berthoz. 'Adaptive Modifications of the Optokinetic Vestibulo-ocular Reflexes in Microgravity.' In *Adaptive Processes in Visual and Oculomotor Systems*, edited by D.S. Zee and E. L Keller, 111–20. New York: Pergamon, 1986.

Wakālat Anbāʾ al-Imārāt. 'Ḥākim al-Shāriqah raʾīsan fakhrīyan lil-Ittiḥād al-ʿArabī li-ʿUlūm al-Faḍāʾ wa-al-Falak.' *Emirates News Agency*, 29 March 2008. http://www.wam.ae/ar/news/emirates/1395235759639.html.

Walgate, Robert. 'French to make Arabsat.' *Nature* 291, no. 5814 (1981): 372.

Waṣfī, Raʾūf. *Al-insān al-ālī al-qātil*. Al-Qāhirah: al-Muʾassah al-ʿArabīyah al-Ḥadīthah [1990s].

———. *Hujūm al-zawāḥif*. Al-Qāhirah: al-Muʾassah al-ʿArabīyah al-Ḥadīthah [1990s].

———. *Al-ḥubb al-mustaḥīl*. Al-Qāhirah: al-Muʾassah al-ʿArabīyah al-Ḥadīthah [1990s].

Wenger, M., F. Ochsenbein, D. Egret, P. Dubois, F. Bonnarel, S. Borde, F. Genova, et al., 'The SIMBAD astronomical database: The CDS reference database for astronomical objects.' *Astronomy and Astrophysics Supplement Series* 143, no. 1 (2000): 9–22.

West, Robert H. Letter to the editor. *Nature* 33, no. 842 (1885): 152.

———. 'Pronunciation of Star Names Derived from the Arabic.' *Popular Astronomy* 2 (1895): 209–14.

———. 'Flight of Birds Across the Moon's Disc.' *Nature* 53, no. 1363 (1895): 131.

———. 'A New Variable in *Sculptor*.' *Astronomical Journal* 16, no. 363 (1896): 23.

———. 'Five New Southern Variables.' *Astronomical Journal* 16, no. 371 (1896): 85–86.

———. 'Two New Southern Variables.' *Astronomical Journal* 16, no. 384 (1896): 211.

———. 'Observations of Southern Variables.' *Astronomical Journal* 17, no. 385 (1896): 3.

———. 'A New Variable of Short Period in *Scorpius*.' *Astronomical Journal* 17, no. 391 (1896): 54.

———. 'Confirmations of Variability.' *Astronomical Journal* 17, no. 395 (1897): 88.

———. 'Confirmations of Variability.' *Astronomical Journal* 18, no. 414 (1898): 46.

Wheatley, N. 'The Planning and Commissioning of a Communications Satellite Earth Station and Its Integration with Existing Telecommunications Systems.' *Radio and Electronic Engineer* 40, no. 5 (1970): 241–54.

Whipple, Fred L., and Salah El-Din Hamid. 'On the Origin of the Taurid Meteor Streams.' *Helwan Observatory Bulletin* 41 (1952): 3–30.

Whole Earth Telescope. 'History.' *Whole Earth Telescope*. Accessed 16 February 2014. http://www.wholeearthtelescope.org/about/history.

Windrem, Robert. 'Iraqi negotiator was Saddam's favorite.' *NBC News*, 24 October 2003. http://www.nbcnews.com/id/3340773/t/iraqi-negotiator-was-saddams-favorite/.

Winter, Donald F. 'The infrared moon: Data, interpretations, and implications.' *Radio Science* 5, no. 2 (1970): 229–40.

Woolley, Richard. 'The RR Lyrae Variable Stars.' *Quarterly Journal of the Royal Astronomical Society* 6, no. 2 (1965): 125–51.

———. 'Riyadh Telescope Committee,' 14 February 1968. RGO 10/780. Royal Greenwich Observatory Archives.

Woolley, Richard, and Khairy Aly. 'Radial-Velocity Observations of RR Lyrae Variables at Kottamia.' *Royal Observatory Bulletins* 114 (1966): E 259–75.

Woolley, Richard, and D.R. Palmer. 'Halo Characteristics of W Virginis Stars.' *Observatory* 87, no. 959 (1967): 173–74.

Worden, Alfred M. *Hello Earth: Greetings from Endeavour*. Los Angeles: Nash, 1974.

Yahia, Mohammed. 'Qatar discovers its second exoplanet.' *Nature Middle East*. 31 May 2012. http://www.nature.com/nmiddleeast/2012/120531/full/nmiddleeast.2012.80.html.

———. 'The false science behind Egyptian army's AIDS and HCV cure.' *House of Wisdom*, 2 March 2014. http://blogs.nature.com/houseofwisdom/2014/03/the-false-science-behind-egyptian-armys-aids-and-hvc-cure.html.

Yamānī, Muḥammad ʿAbduh. *Naẓarāt ʿilmīyah ḥawl ghazw al-faḍāʾ*. Jiddah [c. 1978].

———. *Aqmār ṣināʿīyah ... ghazw jadīd*. [Al-Riyāḍ:] Gulfvision, 1984.

———. 'Aqmār al-faḍāʾ. ghazw thaqāfī jadīd!' *Al-Dawḥah*, no. 9 (September 1984): 9–13.

Zahed, Adnan. 'Saudi University Policy: King Abdulaziz Response.' *Science* 335, no. 6072 (2012): 1040–41.

Zakī, Aḥmad. 'Wahj al-Qamar: Faḍaḥat'hu al-ʿadasāt fa-idhā huwa qiṭʿah min ḥajar.' *Al-ʿArabī*, no. 10 (1 September 1959): 74–81.

Zambrana, Marga. 'AP Interview: Syria's first man in space says European nations should help remove President Bashar Assad from office to bring about an end to the refugee crisis.' *US News & World Report*, 9 March 2016. http://www.usnews.com/news/world/articles/2016-03-09/ap-interview-syrian-cosmonaut-europe-must-help-oust-assad.

Zeinobia. 'Lost in Space: Egypt-Sat 1.' *Egyptian Chronicles*, 23 October 2010. http://egyptianchronicles.blogspot.fr/2010/10/lost-in-space-egypt-sat-1.html.

Zirin, H. 'Coronal line at 4412A.' *The Observatory* 75 (1955): 221–22.

Zohny, Hazem. 'NASA maps groundwater beneath Arabian deserts.' *Nature Middle East*, 13 December 2012. http://www.natureasia.com/en/nmiddleeast/article/10.1038/nmiddleeast.2012.177.

Zol-Faqar, Nour. 'President's scientific adviser delays return to Egypt.' *Cairo Post*, 2 March 2014. http://www.thecairopost.com/news/98069/news/presidents-scientific-adviser-delays-return-to-egypt.

## SECONDARY SOURCES

Abu-Argoub, Ibrahim Ahmad. 'Historical, Political And Technical Development Of ARABSAT.' Doctor of Philosophy dissertation, Northwestern University, 1988.

Abunnasr, Maria. 'The Making of Ras Beirut: A Landscape of Memory for Narratives of Exceptionalism.' Doctor of Philosophy dissertation, University of Massachusetts Amherst, 2013.

———. '"Another Amherst on the Site of Ancient Berytus": Early Impacts of New England on the Ras Beirut Landscape, 1870–1920.' In *One Hundred and Fifty*, edited by Nadia Maria El-Cheikh, Lina Choueiri, and Bilal Orfali, 145–69. Beirut: AUB Press, 2016.

Aiad, A. 'Teaching and Popularizing Astronomy in Egypt and Other Arab Countries.' In *The Teaching of Astronomy*, edited by J.M. Pasachoff and J.R. Percy, 398–99. Cambridge: Cambridge University Press, 1990.

Al-ʿAlāwinah, Aḥmad. *Dhayl al-aʿlām*. Jiddah: Dār al-Manāyirah, 1998.

Al-Azm, Sadik J. 'Islam and the science–religion debates in modern times.' *European Review* 15, no. 3 (2007): 283–95.

Al-Khalili, Jim. *The House of Wisdom: How Arabic Science Saved Ancient Knowledge and Gave Us the Renaissance*. New York: Penguin Press, 2011.

Al-Shārūnī, Yūsuf. 'Al-khayāl al-ʿilmī fī adab Tawfīq al-Ḥakīm.' *Fikr wa-ibdāʿ*, no. 5 (March 2000): 78–89.

———. 'Yūtūbiyā al-khayāl al-ʿilmī fī al-riwāyah al-ʿArabīyah al-muʿāṣirah.' *ʿĀlam al-fikr*, no. 1 (July 2000): 186–211.

Anderson, Betty S. *The American University of Beirut: Arab Nationalism and Liberal Education*. Austin: University of Texas Press, 2011.

Appiah, Kwame Anthony. 'Cosmopolitan Patriots.' *Critical Inquiry* 23, no. 3 (1997): 617–39.

Azoulay, Gérard. 'Diversité et périmètres du patrimoine spatial.' *La revue pour l'histoire du CNRS*, no. 14 (2006).

Bahgat, Gawdat. 'Proliferation of Weapons of Mass Destruction: The Case of Libya.' *International Relations* 22, no. 1 (2008): 105–26.

Bailey, Clinton. 'Bedouin Star-Lore in Sinai and the Negev.' *Bulletin of the School of Oriental and African Studies* 37, no. 3 (1974): 580–96.

Baker, Philip. *The Story of Manned Space Stations: An Introduction*. New York: Springer, 2007.

Balogh, Werner, Levent Canturk, Sergei Chernikov, Takao Doi, Sharafat Gadimova, Hans Haubold, and Victor Kotelnikov. 'The United Nations Programme on Space Applications: Status and Direction for 2010.' *Space Policy* 26, no. 3 (2010): 185–88.

Barbaro, Ada. *La fantascienza nella letteratura araba*. Roma: Carocci, 2013.

Bassi, Joseph P. *A Scientific Peak: How Boulder Became a World Center for Space and Atmospheric Science*. Boston: American Meteorological Society, 2015.

Bermudez Jr, Joseph S. 'Pyramid scheme: Egypt's ballistic missile test and launch facility.' *Jane's Intelligence Review* 22, no. 3 (2010): 48–52.

Brenchley, Frank. *Britain, the Six-Day War and Its Aftermath*. London: I.B.Tauris, 2005.

Brentjes, Sonja. 'The Prison of Categories—"Decline" and Its Company.' In *Islamic Philosophy, Science, Culture and Religion: Studies in Honor of Dimitri Gutas*, edited by Felicitas Opwis and David Reisman, 131–56. Leiden: Brill, 2012.

Burgess, Colin, and Bert Vis. *Interkosmos: The Eastern Bloc's Early Space Program*. Cham: Springer, 2016.

Carus, W. Seth, and Joseph S. Bermudez. 'Iraq's Al-Husayn missile programme, part 1.' *Jane's Soviet Intelligence Review* 2, no. 5 (1990): 204–09.

Crossley, Robert. 'Percival Lowell and the History of Mars.' *Massachusetts Review* 41, no. 3 (2000): 297–318.

D'Afflitto, Isabella Camera. 'Fantascienza in Mauritania. La storia di un uomo nato nel 1034 e morto nel 2055: *Madīnat al-riyāḥ* di Mūsà Wuld Ibnū.' *Oriente Moderno* 16 (77), nos. 2–3 (1997): 331–40.

Dallal, Ahmad [see also El-Dallal, Ahmad Salim]. 'Provost Ahmad Dallal.' *American University of Beirut*. Accessed 23 January 2015. http://www.aub.edu.lb/provost/about/Pages/prov_dallal.aspx.

Danforth, Loring M. *Crossing the Kingdom: Portraits of Saudi Arabia*. Oakland: University of California Press, 2016.

Dorman, John M. 'Missions and Medicine at Amherst: Family Ties to Edward Hitchcock Jr, the Missionary Movement, and the American University of Beirut.' *Journal of American College Health* 59, no. 6 (2011): 489–92.

Doyle, Stephen E. 'Communication Satellites: International Organization for Development and Control.' *California Law Review* 55, no. 2 (1967): 431–48.

Dupont, Anne-Laure. 'Une école missionnaire et étrangère dans la tourmente de la révolution constitutionnelle ottomane.' *Cahiers de la Méditerranée*, no. 75 (2007): 39–57.

El-Ariss, Tarek. 'Let There Be *Nahdah*!' *Cambridge Journal of Postcolonial Literary Inquiry* 2, no. 2 (2015): 260–66.

El-Dallal, Ahmad Salim [see also Dallal, Ahmad]. 'The Astronomical Work of Ṣadr al-Sharīʿah: An Islamic Response to Greek Astronomy.' Doctor of Philosophy dissertation, Columbia University, 1990.

Elshakry, Marwa. 'The Gospel of Science and American Evangelism in Late Ottoman Beirut.' *Past and Present* 196, no. 1 (2007): 173–214.

———. *Reading Darwin in Arabic, 1860–1950*. Chicago: University of Chicago Press, 2013.

Elsheshtawy, Yasser. 'The Prophecy of *Code 46*: Afuera in Dubai, or Our Urban Future.' *Traditional Dwellings and Settlements Review* 22, no. 2 (2011): 19–31.

Ende, Werner. 'Religion, Politik und Literatur in Saudi-Arabien: Der geistesgeschichtliche Hintergrund der heutigen religiösen und kulturpolitischen Situation (III).' *Orient* 23, no. 3 (1982): 378–93.

Evans, Ben. *Tragedy and Triumph in Orbit: The Eighties and Early Nineties*. New York: Springer, 2012.

Farag, Nadia. 'The Lewis Affair and the Fortunes of al-Muqtataf.' *Middle Eastern Studies* 8, no. 1 (1972): 73–83.

Farah, Caesar E. 'Awakening Interest in Western Science & Technology in Ottoman Syria.' In *Transfer of Modern Science & Technology to the Muslim World*, edited by Ekmeleddin İhsanoğlu, 405–18. İstanbul: Research Centre for Islamic History, Art and Culture, 1992.

Freitag, Ulrike. 'Cosmopolitanism in the Middle East as part of global history.' *Programmatic Texts* 4. Berlin: Zentrum Moderner Orient, 2010.

Gaillard, Jacques, and Anne Marie Gaillard. 'Introduction: The International Mobility of Brains: Exodus or Circulation?' *Science, Technology & Society* 2, no. 2 (1997): 195–228.

Gau, Gérard, and Isabelle Sourbes. 'Un système régional de télécommunications spatiales: Arabsat.' *Mappemonde*, no. 1 (1989): 18–20.

Glaß, Dagmar. *Der* Muqtaṭaf *und seine Öffentlichkeit: Aufklärung, Räsonnement und Meinungsstreit in der frühen arabischen Zeitschriftenkommunikation*. Würzburg: Ergon, 2004.

Glass, I.S. 'Early Days at SAAO.' *Monthly Notes of the Astronomical Society of South Africa* 60, no. 1 and 2 (2001): 13–19.

Goldstein, Irving. 'INTELSAT and the Developing World.' *IEEE Transactions on Communications* 24, no. 7 (1976): 742–48.

Gonzalez-Quijano, Yves. 'Aux frontières de la littérature: "un écrivain nommé Mu'ammar al-Qaḏḏāfī".' *Oriente Moderno* 16 (77), nos. 2–3 (1997): 25–35.

Gorman, Alice. 'The archaeology of space exploration.' *Sociological Review* 57 (2009): 132–45.

Grafton, David D. *The Contested Origins of the 1865 Arabic Bible: Contributions to the Nineteenth Century* Nahḍa. Leiden: Brill, 2015.

Guillermier, Pierre, and Serge Koutchmy. *Total Eclipses: Science, Observations, Myths and Legends*. Chichester: Praxis, 1999.

Hanley, Will. 'Grieving Cosmopolitanism in Middle East Studies.' *History Compass* 6, no. 5 (2008): 1346–67.

Harding, Robert C. *Space Policy in Developing Countries: The Search for Security and Development on the Final Frontier*. Abingdon: Routledge, 2013.

Harland, David M. *The Story of the Space Shuttle*. New York: Springer, 2004.

———. *The Story of Space Station Mir*. New York: Springer, 2005.

Ibrahim, Hussein, Abdul Rahim Loulou, and Marwan Koudmani. 'GORS and Its Activities in Syria.' *International Archives of Photogrammetry and Remote Sensing* XXXIII, part B6 (2000): 145–52.

Ibrahim, H., and M. Koudmani. 'A Brief Account of the General Organization of Remote Sensing (GORS) in Syria and Its Activities.' In *New Space Markets*, edited by G. Haskell and M. Rycroft, 345–46. Dordrecht: Springer, 1998.

Inhorn, Marcia C. *Cosmopolitan Conceptions: IVF Sojourns in Global Dubai.* Durham: Duke University Press, 2015.

———. 'Cosmopolitan conceptions in global Dubai? The emiratization of IVF and its consequences.' *Reproductive Biomedicine & Society Online* 2 (2016): 24–31.

Irbah, Abdenour, Toufik Abdelatif, and Hamid Sadsaoud. 'Astronomy in Algeria: Past and Present Developments.' In *Astronomy for Developing Countries*, edited by Alan H. Batten, 171–78. San Francisco: Astronomical Society of the Pacific, 2001.

Jeha, Shafik. *Darwin and the Crisis of 1882 in the Medical Department.* Translated by Sally Kaya. Beirut: American University of Beirut Press, 2004.

Kalder, Daniel. 'Dictator-lit: Gaddafi's surreal gibberish.' *The Guardian*, 24 March 2010, http://www.theguardian.com/books/booksblog/2010/mar/24/dictator-lit-gaddafi.

Kamrava, Mehran. *Qatar: Small State, Big Politics.* Ithaca: Cornell University Press, 2013.

Kandil, Hamdy. 'Le satellite d'Aladin. Le système de communication du satellite arabe.' *Revue Tiers Monde* 28, no. 111 (1987): 659–70.

Karam, John Tofik. 'Philip Hitti, Brazil, and the Diasporic Histories of Area Studies.' *International Journal of Middle East Studies* 46, no. 3 (2014): 451–71.

Karimi Alavi, Mahmoud. 'Contradictions in the Arab Media: The Case of Arabsat.' Doctor of Philosophy thesis, University of Bradford, 2010.

Karp, Aaron. 'Space technology in the Third World: Commercialization and the spread of ballistic missiles.' *Space Policy* 2, no. 2 (1986): 157–68.

Kass, Lee. 'Iran's Space Program: The Next Genie in a Bottle?' *Middle East Review of International Affairs* 10, no. 3 (2006): 15–32.

Kavanaugh, Andrea L. 'Algeria, Morocco, and Tunisia.' In *Telecommunications in Africa*, edited by Eli M. Noam, 13–38. New York: Oxford University Press, 1999.

Kennedy, Nora, Michael Kennedy, David King, and Julio Samsó. 'Edward S. Kennedy (1912–2009).' *Suhayl* 9 (2010): 185–214.

Kerr, Malcolm H. 'Egyptian Foreign Policy and the Revolution.' In *Egypt since the Revolution*, edited by P.J. Vatikiotis, 114–34. London: George Allen & Unwin, 1968.

Koudmani, Marwan. 'A Brief Account of the General Organization of Remote Sensing and Its Activities.' In *Technical Commission VI: Economic, Professional and Educational Aspects of Photogrammetry and Remote Sensing,* edited by Lawrence W. Fritz and James R. Lucas, 157–59. Washington: International Society for Photogrammetry and Remote Sensing, 1992.

Kraidy, Marwan M., and Joe F. Khalil. *Arab Television Industries*. Basingstoke: Palgrave Macmillan, 2009.

Krige, John. 'Embedding the National in the Global: US–French Relationships in Space Science and Rocketry in the 1960s.' In *Science and Technology in the Global Cold War*, edited by Naomi Oreskes and John Krige, 227–50. Cambridge: MIT Press, 2014.

Krige, John, Angelina Long Callahan, and Ashok Maharaj. *NASA in the World: Fifty Years of International Collaboration in Space*. New York: Palgrave Macmillan, 2013.

Lacey, Robert. *Inside the Kingdom: Kings, Clerics, Modernists, Terrorists and the Struggle for Saudi Arabia*. London: Hutchinson, 2009.

Lele, Ajey. *Asian Space Race: Rhetoric or Reality?* Heidelberg: Springer, 2013.

Lewis, Cathleen S. 'Muslims in Space: Observing Religious Rites in a New Environment.' *Astropolitics* 11, no. 1–2 (2013): 108–15.

MacLean, Matthews. 'Time, space and narrative in Emirati museums.' In *Representing the Nation: Heritage, museums, national narratives, and identity in the Arab Gulf States*, edited by Pamela Erskine-Loftus, Victoria Penziner Hightower, and Mariam Ibrahim Al-Mulla, 191–204. Abingdon: Routledge, 2016.

Makdisi, Ussama. 'Reclaiming the Land of the Bible: Missionaries, Secularism, and Evangelical Modernity.' *American Historical Review* 102, no. 3 (1997): 680–713.

Marchisio, Sergio. 'The Evolutionary Stages of the Legal Subcommittee of the United Nations Committee on the Peaceful Uses of Outer Space (COPUOS).' *Journal of Space Law* 31 (2005): 219–42.

Masad, Mohammad. 'Dubai: What Cosmopolitan City?' *ISIM Review* 22, no. 1 (2008): 11–12.

McCray, W. Patrick. *The Visioneers: How a Group of Elite Scientists Pursued Space Colonies, Nanotechnologies, and a Limitless Future*. Princeton: Princeton University Press, 2013.

McDougall, James. 'Dream of Exile, Promise of Home: Language, Education, and Arabism in Algeria.' *International Journal of Middle East Studies* 43, no. 2 (2011): 251–70.

Meloy, John L. 'Arab and Middle Eastern Studies at AUB: Between Local Concerns and Global Pressures.' In *One Hundred and Fifty*, edited by Nadia Maria El-Cheikh, Lina Choueiri, and Bilal Orfali, 85–94. Beirut: AUB Press, 2016.

Misch, Anthony, and William Sheehan. 'A movie of the 1882 Transit of Venus assembled from plates taken at Lick Observatory by David P. Todd.' *Journal of Astronomical Data* 10, no. 4 (2004): 1–9.

Moltz, James Clay. *Asia's Space Race: National Motivations, Regional Rivalries, and International Risks*. New York: Columbia University Press, 2011.

Moulin, Hervé. 'The French–US space research cooperation in the early 1960s.' *Acta Astronautica* 52, no. 8 (2003): 641–48.

Mullane, R. Mike. *Riding Rockets: The Outrageous Tales of a Space Shuttle Astronaut.* New York: Scribner, 2006.

Mumford, George S. 'Dugan, Raymond Smith.' In *Biographical Encyclopedia of Astronomers*, edited by Thomas Hockey, 314–15. New York: Springer, 2007.

Munns, David P.D. *A Single Sky: How an International Community Forged the Science of Radio Astronomy.* Cambridge: MIT Press, 2012.

Navias, Martin S. 'Ballistic missile proliferation in the Middle East.' *Survival* 31, no. 3 (1989): 225–39.

Nielsen, Kai. 'Cosmopolitan Nationalism.' *The Monist* 82, no. 3 (1999): 446–68.

Nizameddin, Talal. *Russia and the Middle East: Towards a New Foreign Policy.* London: Hurst, 1999.

North, Gerald. *Observing the Moon: The modern astronomer's guide.* Cambridge: Cambridge University Press, 2000.

Orton, Karen. 'The desert of the unreal.' *Dazed*, 9 November 2012. http://www.dazeddigital.com/artsandculture/article/15040/1/the-desert-of-the-unreal.

Paikowsky, Deganit, and Isaac Ben Israel. 'Science and technology for national development: The case of Israel's space program.' *Acta Astronautica* 65, no. 9–10 (2009): 1462–70.

Pang, Alex Soojung-Kim. 'The Social Event of the Season: Solar Eclipse Expeditions and Victorian Culture.' *Isis* 84, no. 2 (1993): 252–77.

———. *Empire and the Sun: Victorian Solar Eclipse Expeditions.* Stanford: Stanford University Press, 2002.

Parks, Lisa. *Cultures in Orbit: Satellites and the Televisual.* Durham: Duke University Press, 2005.

Penrose, Stephen B.L. *That They May Have Life: The Story of the American University of Beirut 1866–1941.* New York: Trustees of the American University of Beirut, 1941.

Phillips, Christopher. '*Team Arab*: *al-Jazeera* and the flagging of everyday Arabism during the 2008 Beijing Olympics.' *Nations and Nationalism* 18, no. 3 (2012): 504–26.

Pirard, Theo. 'German Rockets in Africa: The Explosive Heritage of Peenemunde.' *Acta Astronautica* 40, no. 12 (1997): 885–98.

Rinnawi, Khalil. *Instant Nationalism: McArabism, al-Jazeera and Transnational Media in the Arab World.* Lanham: University Press of America, 2006.

Roberts, Greg. 'Paardefontein Satellite Tracking Station 1965–1981.' *Monthly Notes of the Astronomical Society of Southern Africa* 71, no. 7/8 (2012): 165–70.

Rockwell, Trevor Sean. 'Space Propaganda "For All Mankind": Soviet and American Responses to the Cold War, 1957–1977.' Doctor of Philosophy thesis, University of Alberta, 2012.

Sa'di, Lutfi M., George Sarton, and W.T. Van Dyck. 'Al-Hakîm Cornelius Van Alen Van Dyck (1818–1895).' *Isis* 27, no. 1 (1937): 20–45.

Sagdeev, Roald, and Susan Eisenhower. 'United States-Soviet Space Cooperation during the Cold War.' *NASA*, 28 May 2008. http://www.nasa.gov/50th/50th_magazine/coldWarCoOp.html.

Said, Mohamed Kadry. 'Missile proliferation in the Middle East: a regional perspective.' *Disarmament Forum*, no. 2 (2001): 49–61.

Sakr, Naomi. *Satellite Realms: Transnational Television, Globalization and the Middle East*. London: I.B.Tauris, 2001.

Saliba, George. 'Solar Observations at the Maraghah Observatory before 1275: A New Set of Parameters.' *Journal for the History of Astronomy* 16 (1985): 113–22.

———. 'Copernican Astronomy in the Arab East: Theories of the Earth's Motion in the Nineteenth Century.' In *Transfer of Modern Science & Technology to the Muslim World*, edited by Ekmeleddin İhsanoğlu, 145–55. İstanbul: Research Centre for Islamic History, Art and Culture, 1992.

———. *A History of Arabic Astronomy: Planetary Theories During the Golden Age of Islam*. New York: New York University Press, 1994.

———. *Islamic Science and the Making of the European Renaissance*. Cambridge: MIT Press, 2007.

———. 'The Pebble That Became A Fist-Full Rock: On the Continued Importance of Edward Said's Orientalism.' *Logos* 2, no. 3 (2011).

———. 'AUB's Role in the Promotion of the Arabic/Islamic Sciences and Their History.' In *One Hundred and Fifty*, edited by Nadia Maria El-Cheikh, Lina Choueiri, and Bilal Orfali, 213–24. Beirut: AUB Press, 2016.

Sargent, Wallace. 'Itinerant astronomer.' *New Scientist* 114, no. 1561 (21 May 1987): 71.

Satia, Priya. 'Drones: A History from the British Middle East.' *Humanity: An International Journal of Human Rights, Humanitarianism, and Development* 5, no. 1 (2014): 1–31.

Sayah, Edward Nassif. 'The American University of Beirut and Its Educational Activities in Lebanon, 1920–1967.' Doctor of Philosophy dissertation, University of North Texas, 1988.

Schayegh, Cyrus. 'Three Questions for Historians of Science in the Modern Middle East and North Africa.' *International Journal of Middle East Studies* 47, no. 3 (2015): 588–91.

Schiavon, Martina. 'Geodesy and Map-Making in France and Algeria: Contests and Collaborations between Army Officers and Observatory Scientists.' In *The Heavens on Earth: Observatories and Astronomy in Nineteenth-Century Science and Culture*, edited by David Aubin, Charlotte Bigg, and H. Otto Sibum, 199–224. Durham: Duke University Press, 2010.

Schubring, Gert. 'Recent research on institutional history of science and its application to Islamic civilization.' In *Science in Islamic Civilisation*, edited by Ekmeleddin İhsanoğlu and Feza Günergun, 19–36. İstanbul: Research Centre for Islamic History, Art and Culture, 2000.

Scrafford, Robert L. 'A Communications Explosion in Oman.' *IEEE Transactions on Communications* 24, no. 7 (1976): 691–94.

Senn, Martin. 'The Arms-Dynamic Pacemaker: Ballistic-Missile Defense in the Middle East.' *Middle East Policy* 16, no. 4 (2009): 55–67.

Shapir, Yiftah. 'Satellites in the Middle East.' *Strategic Assessment* 7, no. 1 (2004): 22–29.

Shears, Jeremy, and Ashraf Ahmed Shaker. 'Harold Knox-Shaw and the Helwan Observatory.' *Journal of the British Astronomical Association* 125, no. 2 (2015): 80–93.

Sheehan, William, and Anthony Misch. '*Ménage à trois*: David Peck Todd, Mabel Loomis Todd, Austin Dickinson, and the 1882 Transit of Venus.' *Journal for the History of Astronomy* 35 (2004): 123–34.

Siddiqi, Asif A. 'An Asian Space Race: Hype or Reality?' In *South Asia at a Crossroads: Conflict or Cooperation in the Age of Nuclear Weapons, Missile Defense, and Space Rivalries*, edited by Subrata Ghoshroy and Goetz Neuneck, 184–98. Berlin: Nomos, 2010.

Sirrs, Owen L. *Nasser and the Missile Age in the Middle East*. Abingdon: Routledge, 2006.

Stevenson, Angus, ed. *The Oxford Dictionary of English*. Third edition. Oxford: Oxford University Press, 2010.

Stolz, Daniel. 'The Lighthouse and the Observatory: Islam, Authority, and Cultures of Astronomy in Late Ottoman Egypt.' Princeton: Princeton University, 2013.

Tarikhi, Parviz. 'Iran's space programme: Riding high for peace and pride.' *Space Policy* 25, no. 3 (2009): 160–73.

———. *The Iranian Space Endeavor: Ambitions and Reality*. Cham: Springer, 2015.

Telhami, Shibley. 'Power, Legitimacy, and Peace-Making in Arab Coalitions: The New Arabism.' In *Ethnic Conflict and International Politics in the Middle East*, edited by Leonard Binder, 43–60. Gainesville: University Press of Florida, 1999.

Tully, Françoise Le Guet, Jean Davoigneau, Jérôme Lamy, Jérôme de La Noë, Jean-Michel Rousseau, and Hamid Sadsaoud. 'Les traces matérielles de la Carte du Ciel: le cas des observatoires d'Alger et de Bordeaux.' In *La Carte du Ciel: Histoire et actualité d'un projet scientifique international*, edited by Jérôme Lamy, 213–35. Les Ulis: EDP Sciences, 2008.

Tully, Françoise Le Guet, and Hamid Sadsaoud. 'La création de l'observatoire d'Alger.' *La Revue du Musée des arts et métiers* 38 (2003): 26–35.

———. 'Astronomical Heritage Sites: Two Early "Mountain" Observatories on the Mediterranean Coast.' *Monuments and Sites* 18, no. 1 (2015): 204–09.

Tyler, W.S. *History of Amherst College during Its First Half Century*. Springfield: Clark W. Bryan, 1873.

Udías, Agustín. *Searching the Heavens and the Earth: The History of Jesuit Observatories*. Dordrecht: Kluwer, 2003.

University of Chicago Press. *The Chicago Manual of Style*. Sixteenth edition. Chicago: University of Chicago Press, 2010.

Varisco, Daniel. 'Folk Astronomy and the Seasons in the Arabian Gulf.' *Al Ma'thurat Al Sha'biyyah*, no. 19 (July 1990): 7–27.

Väyrynen, Raimo. 'The Arab Organization of Industrialization: A Case Study in the Multinational Production of Arms.' *Current Research on Peace and Violence* 2, no. 2 (1979): 66–79.

Vaz, Jeanne Juliana. 'The Politics of Cold War Cultural Exchange: An Investigation into the Dynamics of Soviet–American Relations During the Post-Stalin Period.' M.Sc. thesis, University of Oxford, 2010.

Völkel, Jan Claudius. *Die Vereinten Nationen im Spiegel führender arabischer Tageszeitungen*. Wiesbaden: VS Verlag für Sozialwissenschaften, 2008.

Von Kries, Wulf. 'The ARABSAT Agreement – Text and Comments.' *Zeitschrift für Luft- und Weltraumrecht* 27 (1978): 179–95.

Watenpaugh, Keith David. 'Cleansing the cosmopolitan city: historicism, journalism and the Arab nation in the post-Ottoman eastern Mediterranean.' *Social History* 30, no. 1 (2005): 1–24.

Westfahl, Gary. *The Spacesuit Film: A History, 1918–1969*. Jefferson: McFarland, 2012.

Wilhelms, Don E. *To a Rocky Moon: A Geologist's History of Lunar Exploration*. Tucson: University of Arizona Press, 1993.

Wilson, O.C. 'Alfred Harrison Joy.' In *Biographical Memoirs*, edited by National Academy of Sciences, 47:225–47. Washington: National Academies Press, 1975.

Ziadat, Adel A. 'Early Reception of Einstein's Relativity in the Arab Periodical Press.' *Annals of Science* 51, no. 1 (1994): 17–35.

# Index

References to illustrations are in italics.

www.ingramcontent.com/pod-product-compliance
Ingram Content Group UK Ltd.
Pitfield, Milton Keynes, MK11 3LW, UK
UKHW020206290926
15470UKWH00001B/22